Green Chemistry Education

ACS SYMPOSIUM SERIES **1011**

Green Chemistry Education

Changing the Course of Chemistry

Paul T. Anastas, Editor
Yale University

Irvin J. Levy, Editor
Gordon College

Kathryn E. Parent, Editor
American Chemical Society

Sponsored by the
ACS Division of Chemical Education, Inc.

American Chemical Society, Washington, DC

ISBN: 978-0-8412-7447-1

The paper used in this publication meets the minimum requirements of American National Standard for Information Sciences—Permanence of Paper for Printed Library Materials, ANSI Z39.48-1984.

Distributed by Oxford University Press

PRINTED IN THE UNITED STATES OF AMERICA

Dedication

To the educators who are
changing the course of chemistry
by advancing green principles and practices

Foreword

The ACS Symposium Series was first published in 1974 to provide a mechanism for publishing symposia quickly in book form. The purpose of the series is to publish timely, comprehensive books developed from ACS sponsored symposia based on current scientific research. Occasionally, books are developed from symposia sponsored by other organizations when the topic is of keen interest to the chemistry audience.

Before agreeing to publish a book, the proposed table of contents is reviewed for appropriate and comprehensive coverage and for interest to the audience. Some papers may be excluded to better focus the book; others may be added to provide comprehensiveness. When appropriate, overview or introductory chapters are added. Drafts of chapters are peer-reviewed prior to final acceptance or rejection, and manuscripts are prepared in camera-ready format.

As a rule, only original research papers and original review papers are included in the volumes. Verbatim reproductions of previously published papers are not accepted.

ACS Books Department

Contents

Indexes

Preface

Any book that strives to describe the *current status* of a dynamic field does so at its peril. So it is with this book. The field of green chemistry is among the most dynamic in science, with a rate of growth and adoption that is not only remarkable but also necessary for our civilization to achieve sustainability. An essential element in the progress of the green chemistry movement is the success of green chemistry education, which provides an understanding of the scientific challenges faced by green chemistry, mastery of the techniques of green chemical practice, insight into the application of the principles of green chemistry, and general advancement of the field. In the absence of this educational component, green chemistry cannot achieve its promise and potential.

Practitioners of green chemistry education, including those represented in the chapters in this book, are showing both the leadership and the diversity of approaches needed to advance green chemistry. Many of the contributors to this volume have taken part in a series of symposia on green chemistry education conducted by the American Chemical Society (ACS) over the course of several ACS national meetings in 2007 and 2008. Current approaches to incorporating green chemistry at all levels and throughout all facets of chemistry education have been documented and discussed. The achievements of both the contributors to this book and the broader chemical education community to date in regard to green chemistry education have been driven by pedagogical insights into new and effective ways to teach chemistry as well as by student interest—often in the form of passionate demand—for exposure to green chemistry.

The only thing more impressive than the accomplishments in the field of green chemistry education thus far is the field's seemingly limitless future potential to accomplish the following:

- attracting students to the field of chemistry who otherwise may never have seen themselves as potential chemists or even as scientists
- providing chemists with an essential skill set that will be needed as the basis of a sustainable world
- bringing a new generation of innovators to engage some of the greatest challenges our society and our civilization face today.

This book has been published in the sincerest hope that it will catalyze the efforts of future educators to build on the work that has already been accomplished and thereby help green chemistry education attain new heights.

In Appreciation:

This book would have not been possible without the extraordinary efforts of my two coeditors, Ms. Kathryn Parent and Professor Irv Levy. In addition, we all thank each of the authors for their contributions that are the essence of this volume.

Paul T. Anastas
Center for Green Chemistry and Green Engineering Chemistry
School of Forestry and Environmental Studies
Chemical Engineering Department
Yale University
New Haven, CT 06520

Chapter 1

Changing the Course of Chemistry

Paul T. Anastas and Evan S. Beach

Center for Green Chemistry and Green Engineering, Chemistry, School of Forestry and Environmental Studies, Chemical Engineering Department, Yale University, New Haven, CT 06520

The education of chemists and all those interested in chemistry is an ever-evolving endeavor to keep up with the latest innovations, discoveries, concepts, perspectives and techniques of the field. One of the most exciting developments in recent years is the development of Green Chemistry – the design of chemical products and processes that reduce or eliminate the generation of toxic substances. This chapter seeks to provide an overview of the approaches to building Green Chemistry into the chemistry curriculum by highlighting some of the outstanding work in the field.

Introduction

The path that the field of chemistry has taken over the course of the past 200 years is one of creativity, innovation, and discovery. It is also a path that we as chemists have followed without fully considering the consequences of either what we have created or the methods and processes we have used to do so. This is largely due to the fact that historically we have had little understanding of the impact of chemicals on human health and the environment. In recent decades, science has dramatically increased our knowledge of the various types of adverse consequences of chemicals. More importantly, it has begun to provide us with a molecular-level understanding of these consequences, thereby allowing us to design our chemical products and transformation processes in order to minimize these adverse consequences. This is the basis of the green chemistry movement, which has been bringing about a wide range of innovations throughout the chemical enterprise.

It is generally accepted that if the still-nascent field of green chemistry is going to have the impact required to allow chemists to play their central role in designing a safer, healthier, and more sustainable world, we must teach the next generation of scientists and educated citizens the fundamental framework of green chemistry. This volume highlights some of the outstanding work from the green chemistry education community that is dedicated to charting this new course.

The Way That Chemistry Has Always Been Taught

In traditional chemical education, technical performance has tended to be the key focus. Chemists are encouraged to achieve clever chemical transformations using any required elements on the periodic table with little regard to hazard or consequence. Elegance in synthesis or chemical processes is rarely described in terms of atom economy, step economy, hazard, amount of waste generated, feedstock use, or other impacts of the chemistry beyond the effects on yield and purity of the target product. Chemists have been taught to accept the fact that handling of explosive, toxic, carcinogenic, or otherwise risky materials is simply part of the nature of the profession. Nonchemists who attend introductory chemistry lectures and labs are unlikely to see any challenge to the general perception of chemistry as the realm of fires, explosions, pollution, and poisons. Reducing risk, which is defined as a function of both hazard and exposure, is almost always achieved by minimizing exposure. In teaching laboratories, this translates to reliance on capital- and upkeep-intensive fume hoods, dependence on personal protective equipment, strict safety training to avoid injury, and adoption of microscale experiments to limit the amounts of dangerous reagents used. To students, the origin of chemical reagents is largely a mystery, and likewise, little thought is given to what happens to the materials poured into the hazardous waste collection jars at the end of every experiment.

Green chemistry education seeks to enhance chemists' understanding of the impacts of their design choices and experiments. Curricula based on the 12 Principles of Green Chemistry (*1*) cast the field of chemistry in an entirely different light. Hazard and waste become recognized as design flaws or, more positively, as opportunities for innovation (*2*). Experiments can be performed in laboratories that are more comfortable and inviting as well as more economical to maintain. Routine handling of inherently safe chemicals reinforces the beneficial impacts and problem-solving potential of the chemical enterprise. Students gain an appreciation for product lifecycles (*3*) and environmental ethics and are empowered to follow their moral inclinations as scientists and engineers (*4*).

Introduction of Green Chemistry as a Field

The idea of green chemistry was initially developed as a response to the Pollution Prevention Act of 1990, which declared that U.S. national policy should eliminate pollution by improved design (including cost-effective changes in products, processes, use of raw materials, and recycling) instead of treatment and disposal. Although the U.S. Environmental Protection Agency (EPA) is known as a regulatory agency, it moved away from the "command and control" or "end of pipe" approach in implementing what would eventually be called its "green chemistry" program. By 1991, the EPA Office of Pollution Prevention and Toxics had launched a research grant program encouraging redesign of existing chemical products and processes to reduce impacts on human health and the environment. The EPA in partnership with the U.S. National Science Foundation (NSF) then proceeded to fund basic research in green chemistry in the early 1990s. The introduction of the annual Presidential Green Chemistry Challenge Awards in 1996 drew attention to both academic and industrial green chemistry success stories. The Awards program and the technologies it highlights are now a cornerstone of the green chemistry educational curriculum. The mid-to-late 1990s saw an increase in the number of international meetings devoted to green chemistry, such as the Gordon Research Conferences on Green Chemistry, and green chemistry networks developed in the United States, the United Kingdom, Spain, and Italy. The 12 Principles of Green Chemistry were published in 1998, providing the new field with a clear set of guidelines for further development (*1*). In 1999, the Royal Society of Chemistry launched its journal *Green Chemistry*. In the last 10 years, national networks have proliferated, special issues devoted to green chemistry have appeared in major journals, and green chemistry concepts have continued to gain traction. A clear sign of this was provided by the citation for the 2005 Nobel Prize for Chemistry awarded to Chauvin, Grubbs, and Schrock, which commended their work as "a great step forward for green chemistry" (*5*).

Green Chemistry in the Classroom

The increasing acceptance of green chemistry in the last decade has been paralleled by the rapid development of green chemistry educational programs around the world, mostly at the undergraduate and graduate levels. Classes are offered by many institutions, ranging from small four-year colleges to major research universities. The first college-level course in green chemistry was taught by Professor Terry Collins at Carnegie Mellon University (CMU) (*6*). Since 1992, the course has been offered to graduate students and advanced undergraduates. In 2008, the course objectives were as follows:

- To understand sustainability ethics as they apply to chemistry and establish the arguments for recognizing "green" criteria.
- To reflect on motives and forces that have entrenched technologies that are obviously or potentially harmful to the environment (*7*).
- To define "green chemistry", place its development in a historical context, introduce the 12 Principles, and study successful examples of green technologies.
- To identify the key challenges facing green chemistry and consider what will be required to solve them (*8*).
- To identify reagents, reactions, and technologies that should be and realistically could be targeted for replacement by green alternatives.
- To understand the history, meaning, and importance of persistent and bioaccumulative pollutants and endocrine disruptors which present major environmental and health threats.
- To become familiar with leading research in green chemistry and the related fields of public health and sustainability science.

The course stresses the link between fundamental chemical concepts and the real-world impacts of chemists' design choices. For example, bond-dissociation energy is taught in the context of ozone depletion, flame retardants, and bleaching technologies; substitution and elimination reactions are discussed in terms of their role in the persistence of organochlorine pollutants in the environment. This link between molecular structure and hazard is discussed in depth in Chapter 8. Sustainability ethics is also a significant component of the course. The material has been developed to show the need for chemists to expand their design capabilities in order to ensure that new chemical products and processes will better consider and incorporate the interests of future generations (*8*, *9*). Another important section of the course focuses on introducing students to endocrine-disrupting chemicals that interfere with cellular processes at environmental concentrations and represent a major design challenge for green chemists. Historical case studies, including cultural reactions to chemistry as well as tensions among industry, the environmental community, and public health professionals, are covered in detail. As the course at CMU moves well into its second decade, it is evolving to take advantage of web technology. Lecture materials will soon be available freely to the global green chemistry community. Furthermore, the course web site is envisioned to become a venue where students, teachers, and experts can interact in problem-solving activities and discussions of state-of-the-art green chemistry research.

At the cutting edge of green chemistry curricula for chemists is the emergence of new molecular design principles (Chapter 8). As green chemistry continues to mature and incorporate learning from a variety of disciplines, courses in the field will provide students with better tools for understanding the molecular bases of acute and chronic toxicity, endocrine-disrupting properties, and physical and global hazards. This will require training in the principles of toxicology, structure–activity relationships, bioavailability, pharmacokinetics, and biological mechanisms of action.

Green chemistry classes for nonmajors are equally important in meeting the goal of equipping the next generation of students to ensure a sustainable future. Many students who may never set foot in a chemistry laboratory nevertheless have a stake in the future of green chemistry. Businesspeople, nontechnical workers in chemical and chemistry-related industries, consumers, and parents concerned about the future of their children all can benefit from being able to make informed decisions about green chemistry issues in their daily lives. Courses in green chemistry and sustainability issues for nonmajors are now appearing at institutions worldwide.

The popularity of green chemistry topics and the drawing power of classes that address them have been noted by many instructors. Positive word-of-mouth referrals and high demand for green chemistry courses are not unusual (Chapter 2). In some cases, students play a key role in bringing green chemistry to campus (Chapter 11). The resonance of green chemistry themes with students can inspire nonmajors to consider the chemistry degree program (Chapter 11) and is poised to improve gender and racial diversity in science and engineering programs (Chapter 9).

Chapter 4 discusses green chemistry course offerings for nonmajors at Simmons College. These classes successfully give students an appreciation of the chemical research process, and students come to realize that environmental problems, when viewed in a different light, become research opportunities. Emphasis on environmental ethics inspires many students to take part in civic action. They raise awareness of green chemistry among the general campus population through high-profile special events and publications.

Yale University has recently launched "Introduction to Green Chemistry", a course for freshman- through senior-level undergraduate nonmajors that accepts students from a wide variety of nonscientific backgrounds. As a course that meets the science core requirements, it contains many green chemistry themes similar to those described above for chemistry majors. However, additional content on the history, trends, and social and cultural aspects of green chemistry is included. The course also requires students to learn about and perform quantitative exercises related to percent yield, atom economy, and introductory toxicology. In addition, students are introduced to green chemistry resources that are highly relevant to daily life. For example, they use the EPA's Toxics Release Inventory (*10*) to find information about pollution in their hometown and access National Institutes of Health databases to learn about the potential hazards of household chemicals. Business case studies are used to show how green chemistry can deliver economic benefits in addition to solving environmental problems. Projects are underway to adapt the material to workshop format for adult education and make the materials freely available on the Internet.

Green Chemistry in the Teaching Laboratory

Green chemistry curriculum for the teaching laboratory was pioneered in the mid-1990s by the chemistry department at the University of Oregon (UO). This work was motivated by the desire to use fewer fume hoods and less-toxic solvents and reagents and to phase out microscale lab equipment. The efforts eventually led to publication of a laboratory manual entitled *Green Organic Chemistry: Strategies, Tools, and Laboratory Experiments* (*11*). As the title indicates, the experiments are designed to replace traditional experiments in organic chemistry laboratories. The new material still covers fundamental organic chemistry reactions and techniques, such as distillation, extraction, melting points, spectral analysis, recrystallization, and thin-layer chromatography, but teaches them in the context of green chemistry topics, such as solvent-free reactions, the utility of molecular modeling calculations to predict the structure of synthetic products, maximum atom economy, catalysis, and waste prevention. As discussed in Chapter 12, this greening of organic laboratory experiments allows chemistry to be performed outside the conventional organic lab setting, allowing a much wider variety of educational institutions to implement laboratory training: if fume hoods are unaffordable or in a very poor state of repair, the green chemistry experiments can still be carried out safely and effectively. A very important side benefit is that the curriculum can be more easily adapted for community colleges and K–12 institutions, making laboratory training available to a higher proportion of younger students, women, and minorities.

The green chemistry efforts of the University of Oregon have had a significant impact due in large part to outreach efforts. Summer workshops hosted at the UO have been very effective in inspiring faculty from other departments to adopt green chemistry laboratory curricula (Chapters 3 and 11 are testimony to this). The UO materials have been designed for easy adoption by others: the strategy has been to suggest replacements for course materials rather than additions, to emphasize fundamental concepts rather than "greenness", to make a business case for implementation of the new curriculum, and to provide a wide range of high-quality materials for faculty to choose from. The UO has also promoted cooperation among green chemistry educators to further improve the quality and quantity of available lab curriculum. The Greener Educational Materials (GEMS) for Chemists Internet database (*12*) developed and hosted by the UO allows researchers to freely exchange laboratory teaching materials. Growth of the GEMS database will undoubtedly play an important role in encouraging more widespread adoption of green chemistry in the laboratory.

Although the green laboratory curriculum is presently most developed for the field of organic chemistry, there are also green experiments for the introductory and analytical chemistry laboratories. For nearly a decade, the

Journal of Chemical Education has been soliciting submissions related specifically to green chemistry; citations for these papers are collected in Chapter 2, and Chapter 5 identifies particular experiments among these that are best suited for the first-year undergraduate curriculum. Chapter 7 discusses the world's first green analytical chemistry laboratory course, offered at Hendrix College for first-year undergraduates. Students gain practical experience with environmental samples, and the experiments are designed to show that the greening of analytical methodology follows many of the 12 Principles of Green Chemistry, including waste prevention, use of safer solvents and auxiliaries, design for energy efficiency, reduction of derivatives, real-time analysis for pollution prevention, and inherently safer chemistry for accident prevention. The class at Hendrix College is connected to a broader effort to disseminate green analytical chemistry knowledge. Other successes of this project include the first review of green analytical chemistry publications (*13*) and application of green "acceptance criteria" to analytical methods in the National Environmental Methods Index (*14*), a free, searchable database containing more than 1,000 analytical, testing, and sampling methods for soil, water, air, and tissues.

Most green laboratory curricula include exercises or discussion to teach students how the procedures have been made green. An interesting approach to emphasizing greenness is practiced by Eötvös University in Budapest, Hungary. The Eötvös lab course requires students to perform conventional lab techniques and the corresponding green replacements side-by-side in order to give the students a tangible appreciation for the waste prevention and risk reduction made possible by green chemistry principles. Students carry out atom economy and E-factor calculations as well as computational modeling of energy balance and lifecycle impacts of both conventional and green methods.

Green Chemistry in Research Training

Most research training in green chemistry takes place under the auspices of traditional chemistry departments, where students pursue graduate or postgraduate research with faculty who have identified themselves as "green". In the last 10 years, such research centers have proliferated around the world, in North and South America, Europe, Asia, Africa, and Oceania (Figure 1). In many cases, government funding has helped establish green chemistry centers, such as NSF centers in the United States, China's Key Labs in Green Chemistry, and Denmark's Center for Green Chemistry and Sustainability. Many green chemistry networks originated through interuniversity collaborations (such as Italy's INCA). Since the late 1990s, the Green Chemistry Institute (GCI), now a branch of the American Chemical Society (ACS), has been helping to establish international green chemistry chapters whose members promote both research and education.

Degree programs specifically in green chemistry are available but still rare. The first Ph.D. program in green chemistry was established at the University of Massachusetts-Boston by Professor John Warner. Amy Cannon, the first student to enroll, graduated from the program in 2005 and now directs Beyond Benign, a nonprofit green chemistry education organization (Chapter 12). A similar Ph.D. program was introduced by Warner at the University of Massachusetts-Lowell, where in addition to an introductory green chemistry course, Ph.D. students can take electives in Toxicology, Sustainable Materials Design, and Environmental Law and Policy. The University of York has been offering a master's-level course since 2001. Initially designated as a Master of Research in Clean Chemical Technology, the program has evolved into a Master of Science in Green Chemistry and Sustainable Industrial Technology. The curriculum is centered on six-month research projects to develop environmentally benign products and processes in partnerships with industry. Monash University in Melbourne, Australia, also offers an M.S. course in green chemistry. In 2007, Cambridge College announced that it would be the first campus in the United States to offer B.S. and M.S. degrees in green chemistry.

Green Chemistry Summer Schools and Training Workshops

A crucial factor in the ability of green chemistry to take root in new institutions around the world has been the success of green chemistry summer schools and training workshops. Many of the programs were designed for younger scientists, from the undergraduate to the junior-faculty level. One of the oldest ongoing series is the Postgraduate Summer School on Green Chemistry, initiated by INCA in 1998. This one-week program in Venice, Italy, has targeted European scientists from academia and industry under the age of 35. The event features expert speakers, tutorial sessions, a poster session with awards, and interactive discussions on green chemistry topics such as catalysis, alternative solvents, green reagents, and research policy (*15*). As the summer school moves into its second decade, the applicant pool has grown significantly.

As green chemistry became more entrenched at the beginning of the 21st century, summer schools and workshops were often intended to introduce green chemistry to a more diverse audience. Los Alamos National Laboratory hosted a two-week workshop in 2001 for 40 young scientists from developing countries. The ACS GCI helped establish an annual summer-school program that encourages graduate students and postdoctoral researchers in the Americas to establish working relationships. The venue changes every summer, with the United States, Canada, Uruguay, and Mexico having served as hosts to date. The International Union of Pure and Applied Chemistry (IUPAC)-sponsored CHEMRAWN XIV conference in 2001, entitled "Toward Environmentally Benign Processes and Products", led to funding for the GCI-administered Developing and Emerging Nations Grants Program, through which grants were

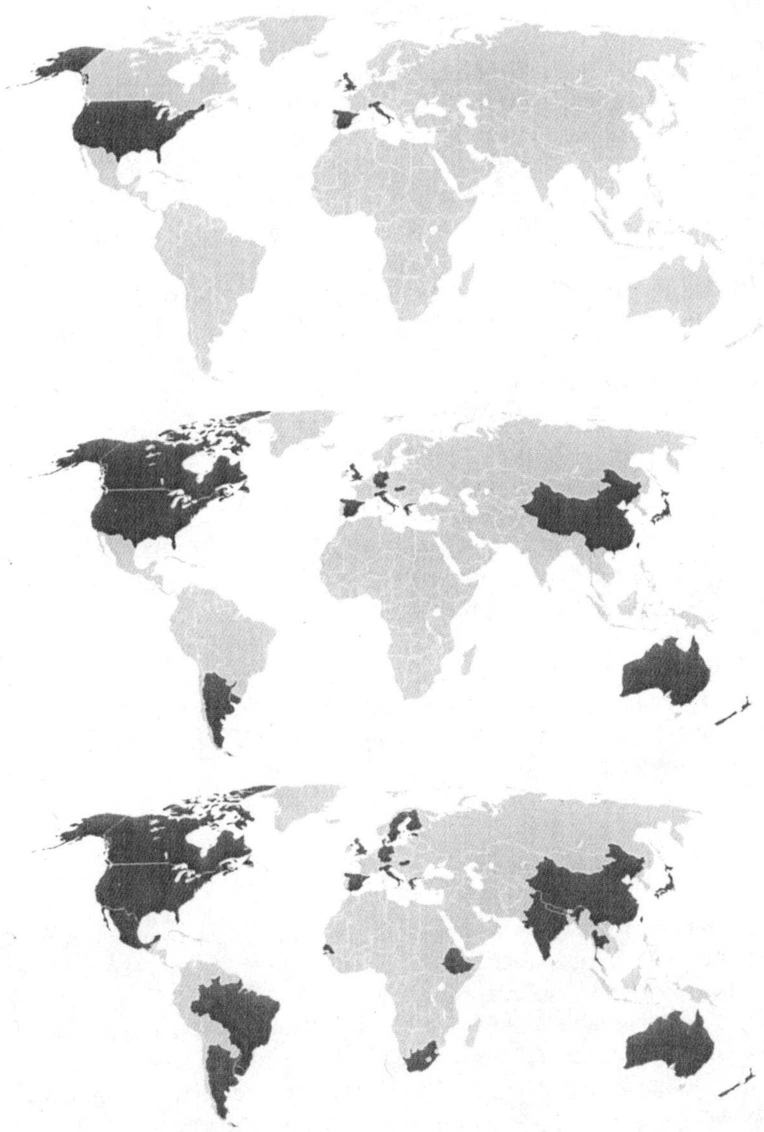

Figure 1. Growth in global green chemistry research and education. (top) In the early 1990s, green chemistry efforts were focused mainly in the United States, Italy, and the United Kingdom. (middle) The late 1990s and early 2000s saw green chemistry networks and GCI chapters make inroads in South America, Asia, and Oceania. (bottom) As of 2008, green chemistry initiatives are underway in Africa, and gains on virtually every continent have been made. (See page 1 of color insert.)

awarded to support educational workshops and training in Ethiopia, Nigeria, the Middle East, India, Nepal, and Hungary. Funding was also provided for translation of educational materials into Spanish and Portuguese.

One of the programs funded was a two-day green chemistry training workshop in Thailand. This event drew students and faculty from Australia, India, Indonesia, Malaysia, and Vietnam. In 2008, India hosted the Third Indo-US Workshop on Green Chemistry. Among the activities were demonstrations of green chemistry laboratory experiments and discussions of methods for developing green chemistry workbooks and problem sets. The inaugural session was attended by India's Union Minister of Mines, Dr. T. Subbarami Reddy, who later sent a letter of request to the Prime Minister to request that green chemistry be prioritized (*16*). The potential benefits that such prioritization could bring to both green chemistry policy and education in India are tremendous, testifying to the power of workshops to bring about real change.

In Africa, Italy's INCA consortium organized the Italian–North African Workshop on Sustainable Chemistry in 2002 to promote green chemistry education in Italy, Algeria, Morocco, Tunisia, and Egypt. Ethiopia has recently emerged as a significant green chemistry center: as of 2007, three annual workshops have already been held there. These had a very practical orientation, with material geared toward the greening of local industries, such as the biofuels, textiles, and food industries. The 2007 workshop led to the organization of smaller regional workshops (*17*). This "training the trainer" model, in which experienced green chemists teach other educators how to design green chemistry education workshops of their own, is among the most efficient and effective ways to disseminate green chemistry educational materials.

Green Chemistry Curriculum Materials

This section provides an annotated list of some of the best-known books available for use in green chemistry lecture and laboratory courses, and additional supplementary material is discussed throughout this volume. Many of the materials are available at low cost, and some can be downloaded at no cost from the GEMS database (*12*) mentioned above. While the books listed in this section are designed specifically for green chemistry courses, it is the goal of many educators to make green material integral to the traditional chemistry curriculum from the introductory level onward, so that ultimately there will be no need to consider green chemistry as a standalone field or special topic—ideally, green chemistry will become standard practice, and we will only have to talk about "chemistry" (Chapters 5 and 6). Integration of green chemistry into general and upper-level chemistry textbooks for undergraduate chemistry majors is in the very early stages of development. The strongest inroads made by green chemistry into mainstream textbooks to date can be found in introductory texts for undergraduate nonmajors. This is discussed in detail in Chapter 6, where two textbooks have been cited as exemplary: *Chemistry for Changing Times*, 11th ed., by Hill and Kolb (*18*) and *Chemistry in Context*, 5th ed., by Eubanks et al. (*19*). The potential of green chemistry to replace traditional material instead of supplementing it or standing alone is further discussed in Chapter 5, which also suggests strategies for overcoming resistance to the adoption of a new green curriculum.

In the annotated lists of green chemistry books given below, the student levels for which the books are appropriate are indicated by the following abbreviations: HS, high school; UN, undergraduate nonmajor; UM, undergraduate major; GR, graduate.

Books for Green Chemistry Lecture Courses

- *Green Chemistry: Theory and Practice* by Anastas and Warner (*1*) [UN, UM, GR]. This book is the original source of the 12 Principles of Green Chemistry and provides illustrations of each of the principles. It includes case studies of greened technologies and highlights of green chemistry research up to 1998. Student exercises for each chapter are also included. See ref *20* for a book review that discusses the use of this book in education.
- *Real-World Cases in Green Chemistry* by Cann and Connelly (*21*) [UN, UM, GR]. This book discusses selected Presidential Green Chemistry Challenge Award winners, covering a broad range of chemistry topics. It has been used in many green chemistry lecture courses to highlight some of the most successful applications of the 12 Principles of Green Chemistry and their relevance to everyday life (*e.g.*, see Chapter 6).
- *Introduction to Green Chemistry* by Matlack (*22*) [UM, GR]. This textbook contains hands-on activities and more than 5,000 references.
- *Green Chemistry: An Introductory Text* by Lancaster (*23*) [UM, GR]. This textbook includes some green engineering material and addresses policy issues. Discussion questions are provided at the end of each chapter. See ref *24* for a book review.
- *Going Green: Introducing Green Chemistry into the Curriculum*, edited by Parent and Kirchhoff (*25*) [UN, UM]. Designed for faculty members, this book provides an overview of green chemistry and sustainability issues and goals and gives examples of how other educators have introduced those themes into the chemistry curriculum.
- *Green Chemistry and the Ten Commandments of Sustainability*, 2nd ed., by Manahan (*26*) [HS, UN]. This book describes the design of safer chemicals and the impacts of chemistry on air, land, and water. It introduces industrial ecology and discusses renewable feedstocks and alternative energy sources.

Books for Green Chemistry Laboratory Courses

- *Greener Approaches to Undergraduate Chemistry Experiments*, edited by Kirchhoff and Ryan (*27*) [UN, UM]. This book is a compilation of 14 organic chemistry laboratory experiments, including pre- and post-laboratory assignments and discussions of how the experiments were designed to be green. It is freely available through the GEMS database (*12*).
- *Introduction to Green Chemistry*, edited by Ryan and Tinnesand (*28*) [HS]. This teaching manual contains a collection of laboratory activities that illustrate the 12 Principles of Green Chemistry. For a book review, see ref *29*.
- *Green Organic Chemistry: Strategies, Tools, and Laboratory Experiments* by Doxsee and Hutchison (*11*) [UN, UM]. This book provides green experiments for teaching common organic chemistry laboratory principles, such as solvent-free reactions, molecular modeling, atom economy, catalysis, and waste prevention.

Green Chemistry Tools and Databases

The Internet has become an increasingly important medium for publishing and sharing green chemistry educational materials, and as Internet access in developing countries continues to improve, web-based educational materials are likely to be more easily accessible and more widely adopted than textbooks. As of 2008, numerous teaching materials, tools, and databases for students and teachers are already available at no cost. These include the following:

- The GEMs database (*12*), developed by the University of Oregon. The GEMs database is perhaps the best web-based source of educational laboratory materials. It includes procedures compiled by the UO and the ACS, all of the experiments contained in Doxsee and Hutchison's organic chemistry laboratory manual (*11*), and contributions previously published in the literature (*e.g.*, by the *Journal of Chemical Education* and *Green Chemistry*). The database also includes modules developed by the University of Scranton's "Greening Across the Chemistry Curriculum" program (*30*) as well as unpublished green chemistry educational materials submitted by chemical educators. The GEMs database allows searching by green chemistry principles, educational level, laboratory techniques, and fundamental chemistry concepts. The website also provides message boards for faculty to discuss tips for adopting the materials and share comments about how the experiments may be improved.
- Resources from the Green Chemistry Institute of the American Chemical Society. The GCI has created a web site offering freely downloadable green

chemistry activities and experiments geared toward all levels of education from elementary school through graduate study (*31*). The GCI also maintains the Green Chemistry Resource Exchange (*32*). Although this database is designed more for technical use than for educational use, it nevertheless serves as a useful resource for green chemistry educators, as it contains information about many Presidential Green Chemistry Challenge Award winners and other successful green technologies and is searchable by industry, applications, and green chemistry targets (for example, use of renewable resources or design of safer chemicals).

- The Green Chemistry Expert System (GCES) (*33*), designed by the EPA Green Chemistry Program. GCES is a freely downloadable computer program developed in 1999 for computers running the Microsoft Windows operating system. It is built around a database of more than 200 green chemistry publications, including journal articles, patents, news items, and product brochures, covering topics such as green solvents and reagents, green chemical processing, green synthesis and manufacturing, and alternative feedstocks. The software contains various modules for browsing and searching this data (Figure 2). For example, the Green Synthetic Reactions module allows keyword searching and provides information on whether a technology is in the research-and-development, pilot-plant, or commercial stage. Although the GCES has not been updated in recent years, this database is still a useful tool for exploring early successes in green chemistry and learning about green technologies that have been implemented on the commercial scale.
- The Green Chemistry Assistant (GCA), developed by St. Olaf College in collaboration with the EPA. The GCA is an enhancement of the Synthetic Methodology Assessment for Reduction Techniques (SMART) module of the GCES. SMART is used for categorizing and (when possible) quantifying the greenness of a chemical process in terms of hazards, feedstocks, and rudimentary life-cycle assessment. This portion of the GCES software has since been greatly extended and made significantly more user-friendly by Professor Robert Hansen and co-workers at St. Olaf College, resulting in the GCA web application (*34*). The GCA user enters the identities and amounts of feedstocks, products, solvents, and auxiliary materials used in an experiment, and the application calculates the resulting atom economy, yield, and excesses and also helps summarize information about solvent usage and safety. The results can be converted into charts that easily allow the user to identify inefficiencies and other problems associated with a process (Figure 3).

To sustain the development of new green chemistry curriculum and tools, a network of faculty has been formed to promote a collaborative approach to preparation of new materials. The Green Chemistry Education Network (GCEdNet) (*35*) began in 2004 with five "ambassador" faculty members located

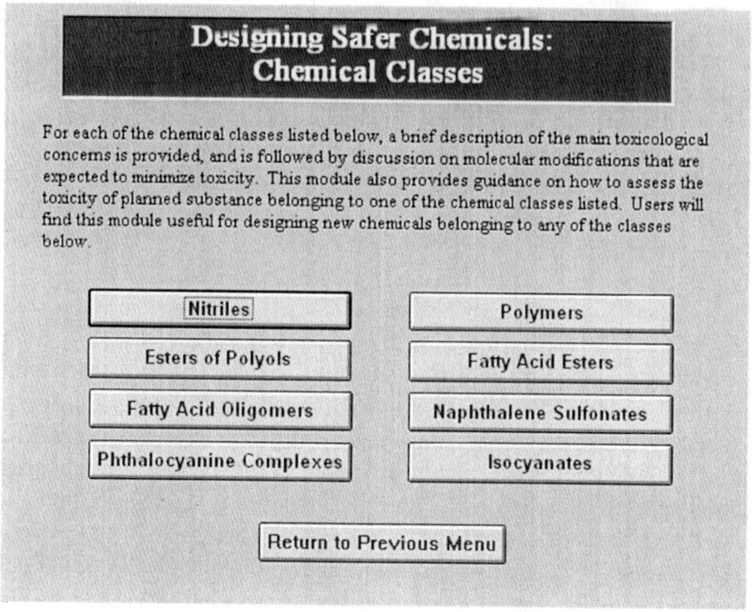

Figure 2. The Designing Safer Chemicals module of EPA's Green Chemistry Expert System. (See page 2 of color insert.)

in three states. These ambassadors coordinate the development of curriculum for universities, colleges, community colleges, and high schools. As of 2008, GCEdNet has grown to include 20 active faculty members in six states. While the activities are currently concentrated in the United States, the network welcomes international participation. Surveys of green chemistry educators have shown that most faculty develop a significant proportion of the curriculum used in their individual courses (Chapter 2). The peer-led, cooperative nature of GCEdNet is one way to save faculty time and resources by avoiding redundant labor. It is also expected to improve access to green chemistry educational materials for teachers at every level.

Green Chemistry for Younger Students

Preparing the next generation of green chemists depends not only on training undergraduate and graduate chemistry majors but also on reaching out to younger students. At the K–12 level, green chemistry education is an excellent way to attract bright students to the chemistry profession. A variety of green chemistry programs exist for elementary and high-school education. Many are based on outreach efforts by undergraduate and graduate students:

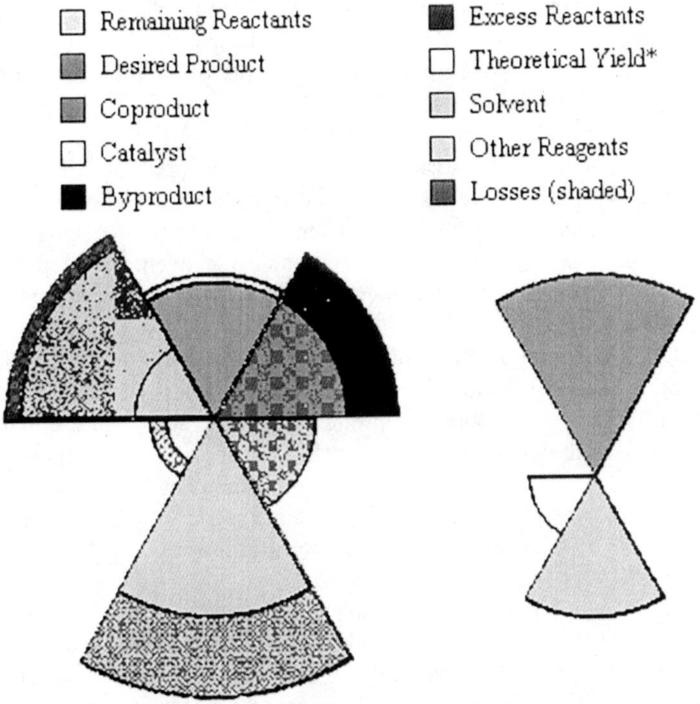

Figure 3. Visualizing atom economy, process efficiency, and waste generation using the Green Chemistry Assistant web application developed by St. Olaf College. The diagram on the left shows a process that uses excess reagents, generates coproducts and byproducts, has poor solvent and catalyst recovery, and involves a reaction that does not go to completion. In contrast, the diagram on the right depicts a 100% atom-economical reaction using small amounts of solvent and catalyst that are completely recovered. Such graphical representations help the user quickly identify problem areas in a chemical process. *(See page 3 of color insert.)*

- The nonprofit organization Beyond Benign promotes service-learning projects, encouraging undergraduate students to enhance their own understanding of green chemistry by creating curriculum for younger students (Chapter 12). The undergraduates interact with K–12 students, highlighting the power of green chemistry to innovate and solve problems and giving a positive impression of chemistry in general. The Beyond Benign program has already led to undergraduate-designed teaching materials for the eighth-grade level.
- Similarly, a program at Gordon College encourages green chemistry students to engage in outreach to high schools, K–12 educators, and home-

schooled children (Chapter 11). A key feature of the Gordon program is the effort to ensure that student-developed material endures well into the future.

- The Clean Technology Group at the University of Nottingham has devised presentations for elementary-school students, again focusing on how chemists can make a positive impact on daily life. Nottingham funds a "Public Awareness Scientist" position at the university to coordinate this effort; this person actively tracks green chemistry developments and finds ways to communicate them to non-scientific audiences (*36*). One method is to encourage Ph.D. students to visit secondary schools on a regular basis to give the younger students an understanding of how green projects develop over time.
- The ACS encourages university students to perform community outreach through its undergraduate Student Affiliate Chapters. Accreditation as a "green" chapter is achieved through activities such as teaching green chemistry in the community (*e.g.*, at local schools and industries). Green chapters also promote green chemistry on campus by hosting speakers and conducting workshops and within chemistry departments by designing greener lab experiments and sponsoring poster sessions.

The ACS, in addition to offering a selection of downloadable green chemistry activities for K–12 classrooms, hosts annual workshops for high-school teachers. These three-day, all-expenses-paid courses train teachers how to use green chemistry educational resources, conduct experiments and demonstrations, and integrate green chemistry principles into the existing science curriculum. Pfizer, Inc., a member of the ACS GCI Pharmaceutical Roundtable, has posted middle-school-level green chemistry curriculum materials in both English and Spanish on its web site (*37*). The lessons include hands-on activities and cover topics such as risk, waste, parts-per-million toxicity, energy efficiency, and material lifecycles.

Conclusion

This review of the excellent work that has been done in changing the course of chemical education to include and integrate green chemistry represents only the tip of the iceberg both of what is currently taking place and what needs to take place in the future. While this chapter does not claim to be comprehensive, it does provide an overview of some of the approaches being considered. The green chemistry education work developing all over the world in countries that have different environmental, educational, economic, cultural, historical, and social circumstances demonstrates that the applicability of the field of green chemistry is as broad as that of the field of chemistry itself.

References and Notes

1. Anastas, P. T.; Warner, J. C. *Green Chemistry: Theory and Practice*; Oxford University Press: Oxford, U.K.,1998.
2. Horvath, I. T.; Anastas, P. T. Innovations and Green Chemistry. *Chem. Rev.* **2007**, *107*, 2169-2173.
3. Lankey, R. L.; Anastas, P. T. Life-Cycle Approaches for Assessing Green Chemistry Technologies. *Ind. Eng. Chem. Res.* **2002**, *41*, 4498-4502.
4. McDonough, W.; Braungart, M.; Anastas, P. T.; Zimmerman, J. B. Applying the Principles of Green Engineering to Cradle-to-Cradle Design. *Environ. Sci. Technol.* **2003**, *37*, 434A-441A.
5. Press Release: The Nobel Prize in Chemistry 2005. http://nobelprize.org/nobel_prizes/chemistry/laureates/2005/press.html (accessed Jan 10, 2008).
6. Collins, T. J. Introducing Green Chemistry in Teaching and Research. *J. Chem. Educ.* **1995**, *72*, 965-966.
7. Collins, T. J. In *Macmillan Encyclopedia of Chemistry*; Simon and Schuster Macmillan: New York, 1997.
8. Collins, T. Essays on Science and Society: Toward Sustainable Chemistry. *Science* **2001**, *291*, 48-49.
9. Collins, T. The Importance of Sustainability Ethics, Toxicity and Ecotoxicity in Chemical Education and Research. *Green Chem.* **2003**, *5*, G51-G52.
10. U.S. Environmental Protection Agency. Toxics Release Inventory (TRI) Program Web Site. http://www.epa.gov/tri (accessed Jan 10, 2008).
11. Doxsee, K. M.; Hutchison, J. E. *Green Organic Chemistry: Strategies, Tools, and Laboratory Experiments*; Thomson Brooks/Cole: Belmont, CA, 2003.
12. The Greener Education Materials (GEMs) for Chemists Database. http://greenchem.uoregon.edu/gems.html (accessed Jan 10, 2008).
13. Keith, L. H.; Gron, L. U.; Young, J. L. Green Analytical Methodologies. *Chem. Rev.* **2007**, *107*, 2695-2708.
14. National Environmental Methods Index. http://www.nemi.gov (accessed Jan 10, 2008).
15. Perosa, A. First Postgraduate Summer School on Green Chemistry, Venice, Italy. *Green Chem.* **1999**, *1*, G25-G27.
16. Dr. R. K. Sharma, Green Chemistry Network Centre, Department of Chemistry, University of Delhi, Delhi, India. Personal communication, 2008.
17. Chebude, Y.; Yilma, D.; Licence, P.; Poliakoff, M. Green Chemistry in Ethiopia: The Third Annual Workshop. *Green Chem.* **2007**, *9*, 822.
18. Hill, J. W.; Kolb, D. K. *Chemistry for Changing Times*, 11th ed.; Prentice Hall: Upper Saddle River, NJ, 2007.

19. Eubanks, L. P.; Middlecamp, C. H.; Pienta, N. J.; Heltzel, C. E.; Weaver, G. C. *Chemistry in Context*, 5th ed.; McGraw-Hill: New York, 2006.
20. Leitner, W. Toward Benign Ends. *Science* **1999**, *284*, 1780-1781.
21. Cann, M. C.; Connelly, M. E. *Real-World Cases in Green Chemistry*; American Chemical Society: Washington, DC, 2000.
22. Matlack, A. *Introduction to Green Chemistry*; Marcel Dekker: New York, 2001.
23. Lancaster, M. *Green Chemistry: An Introductory Text*; Royal Society of Chemistry: London, 2002.
24. Rosan, A. M. Green Chemistry: An Introductory Text (Mike Lancaster). *J. Chem. Educ.* **2003**, *80*, 1141-1142.
25. *Going Green: Introducing Green Chemistry into the Curriculum*; Parent, K., Kirchhoff, M., Eds.; American Chemical Society: Washington, DC, 2004.
26. Manahan, S. *Green Chemistry and the Ten Commandments of Sustainability*, 2nd ed.; ChemChar Research, Inc.: Columbia, MO, 2005.
27. *Greener Approaches to Undergraduate Chemistry Experiments*; Kirchhoff, M., Ryan, M. A., Eds.; American Chemical Society: Washington, DC, 2002.
28. *Introduction to Green Chemistry*; Ryan, M. A., Tinnesand, M., Eds.; American Chemical Society: Washington, DC, 2002.
29. Conover, W. Introduction to Green Chemistry (Mary Ann Ryan and Michael Tinnesand). *J. Chem. Educ.* **2003**, *80*, 268.
30. Greening Across the Chemistry Curriculum Home Page. http://academic.scranton.edu/faculty/cannm1/dreyfusmodules.html (accessed Jan 10, 2008).
31. The ACS Green Chemistry Institute. Green Chemistry Educational Resources Web Page. http://portal.acs.org/portal/Navigate?nodeid=1444 (accessed Jan 10, 2008).
32. The ACS Green Chemistry Institute. Green Chemistry Resource Exchange. http://www.greenchemex.org (accessed Jan 10, 2008).
33. U.S. Environmental Protection Agency. Green Chemistry Expert System (GCES) version 0.99 Home Jan Page. http://www.epa.gov/greenchemistry/pubs/gces.html (accessed Jan 10, 2008).
34. Hanson, R.; Campbell, P.; Christianson, A.; Klingshirn, M.; Engler, R. Green Chemistry Assistant. http://fusion.stolaf.edu/gca (accessed Jan 10, 2008).
35. Green Chemistry Education Network Home Page. http://www.gcednet.org/ (accessed Jan 10, 2008).
36. Hager, Y. Careers: The Outreach Bug. *Chem. World* **2006**, *3*.
37. Pfizer, Inc. Recipe for Sustainable Science: An Introduction to Green Chemistry in the Middle School Web Page. http://www.pfizer.com/responsibility/education/green_curriculum.jsp (accessed Jan 10, 2008).

Chapter 2

Using Green Chemistry to Enhance Faculty Professional Development Opportunities

Margaret E. Kerr[1] and David M. Brown[2]

[1]Department of Chemistry, Worcester State College, Worcester, MA 01602
[2]Department of Chemistry, Davidson College, Davidson, NC 28035

Of the plethora of benefits that derive from practicing green chemistry, one that is not often considered, or at least discussed, is its application toward enhancing the professional development of faculty as they advance through the ranks. Opportunities within the areas of teaching, scholarly activities (research and related), and service (both community and institutional) abound as the field advances rapidly. Within the context of developing new green chemistry educational materials, herein is presented a discussion of multiple professional development opportunities taken from both the authors' personal experiences as well as from the numerous contributions of others in the field.

Introduction

It is widely acknowledged by members of the green chemistry community that the introduction and study of green chemistry into the collegiate curriculum are vital to the success of the overall movement. According to Professor Walter Leitner, scientific editor of the journal *Green Chemistry*, "As the principles of green chemistry and the concepts of sustainability in chemical manufacturing are becoming part of the explicit corporate policy and aims in the chemical industry, there is a rapidly growing need for the education of chemists in the field." *(1)* The question that arises is "Who will be developing the materials and curriculum for this to happen?" Any type of new scholarship and implementation of new teaching materials by individual faculty members must be rewarded by their home institutions in order for sustainable programs to develop. It is imperative for those interested in the development of green chemistry education materials to expect that their contributions be considered a part of any scholarly portfolio.

We have separated this chapter into three sections: Teaching, Scholarly Activity, and Service. These three topics are the primary components of most tenure and promotion decisions at any institution and, therefore, are useful to address individually. We have given primary focus to teaching and scholarly activity, with a smaller focus on service. We have also tried to incorporate as many diverse examples as possible to recognize that the term "professional development" has different meanings depending on the type of institution involved. All of the examples listed in this chapter have been considered professional development by the institution where they were developed.

Since the publication in 1998 of the seminal work of Anastas and Warner *(2)*, interest in green chemistry has dramatically increased. A particular focus on green chemistry education has given rise to new research programs and opportunities for professional development. Groups such as the Green Chemistry Education Network (GCEdNet) *(3)* have expanded green chemistry educational materials development and dissemination throughout the nation. This group has fostered professionalism and collaborative work arrangements with its members and has encouraged the expansion of membership by organizing meetings nationally and regionally. A major focus has been to provide members with ways of developing programs that will expand the amount of research and development of new education materials while ensuring that the work done will be taken seriously and rewarded by the scholar's home institution.

A questionnaire was sent to members of two green chemistry databases in May of 2007. This questionnaire was designed to find out how the practice of green chemistry has affected the professional development opportunities for faculty members at different institutions. A total of forty-one people responded to the questionnaire. The respondents were primarily from B.S. granting departments (56%), with smaller numbers from M.S. and Ph.D. granting departments (15% and 17%, respectively). A small group of community college and high school faculty also responded. The institutions were evenly split between public and private. The majority of respondents had been incorporating green chemistry in their curriculum for more than two years (68%). Only one respondent has incorporated no green chemistry into the curriculum.

The majority of activity in incorporating green chemistry curriculum has been in the required laboratory for science majors. The next highest has been used in lecture courses, also required for science majors. Significant work has gone into integrating material into courses for non-majors as well.

Interesting questionnaire results showed varied responses for the level of development that has occurred for individual courses. The majority of respondents have developed 25-50% of the green material for their lecture and laboratory courses. A smaller portion of people have developed 75-100% of the materials for their courses, and a similar number of people have done no development work. The results of this particular question set demonstrate not only that the material already published and available is useful for faculty members, but also that there is a significant body of people who are working to create more curriculum.

Criteria that are used in professional advancement decisions include publication in peer-reviewed journals and presentations at professional meetings. Forty-nine percent of respondents have disseminated materials beyond their own classrooms in a variety of ways. Significant numbers of those who have done so have published work in peer-reviewed journals (32%), and almost twice as many gave oral presentations of their work at professional meetings (63%). An important component of green chemistry education is the involvement of students in the research programs of faculty members. Material was also distributed by students at professional meetings, primarily as poster presentations.

Over eighty percent of respondents said that the implementation of green chemistry into the curriculum has helped them professionally. The distribution of responses as to how it has helped was fairly evenly divided between the hiring process, pre-tenure or yearly review, tenure, promotion, and post-tenure review. Additionally, the implementation of green chemistry into the curriculum aided fifty percent of respondents with their grant seeking activities. The majority of grant funding has been from individual institutions, with foundation grants second. A much smaller percentage (20%) of people have received government grants. There was not a question that addressed success rates for green chemistry grant applications to government agencies.

The following sections will review work that has been accomplished by members of the rapidly expanding green chemistry community and will serve as a resource for others who wish to incorporate green chemistry into their curricula.

Teaching

Scant educational supporting materials were available in the early days of the paradigm shift toward incorporating green chemistry into the traditional American chemistry curriculum. Fortunately, during the past few years, this has changed rather dramatically. Nowadays, considerably more supporting resources are available in the form of textbooks, reference works and monographs, laboratory manuals, web sites, a video, and journal articles focused on green chemistry education. Although this is not an exhaustively inclusive list, below are listed some currently available key educational resources focused solely on green chemistry:

Textbooks

- Anastas and Warner (1998) *(2)*
- Matlack (2001) *(4)*
- Lancaster (2002) *(5)*
- Manahan (2005) *(6)*

Reference works

- Stevens (2001) *(7)*
- Clark and Macquarrie (2002) *(8)*
- Rogers and Seddon (2003) *(9)*
- Koichi (2005) *(10)*
- Srivastava (2005) *(11)*
- Rogers and Seddon (2006) *(12)*
- Sheldon, Arends, *et al.* (2007) *(13)*

Laboratory manuals

- Kirchhoff and Ryan (2002) *(14)*
- Doxsee and Hutchison (2004) *(15)*

ACS booklets and other supplemental support materials

- Cann and Connelly (2000) *(16)*
- Ryan and Tinnesand (2002) *(17)*
- Parent, Kirchhoff, and Godby (2004) *(18)*

Web sites

- American Chemical Society (ACS) Green Chemistry Institute *(19)*
- United States Environmental Protection Agency (US EPA) *(20)*
- Green Chemistry Network (GCN) *(21)*
- Greener Education Materials for Chemists (GEMs) *(22)*
- Green Chemistry Education Network (GCEdNet) *(3)*

Video

- Weise, E. "Green Chemistry: Innovations for a Cleaner World" (2000) *(23)*

These works constitute substantial research projects and have been accomplished over relatively long periods of time. Many faculty members have not had the time or opportunity to create works of this magnitude, but have been developing materials in the context of their own courses.

Development of a Green Chemistry Course at Davidson College

In 2002, through inspiration received by attending one of the excellent, hands-on "Green Chemistry in Education" Workshops offered at the University of Oregon, David Brown set out to design one of the first "stand-alone" courses in green chemistry for undergraduates to be offered at Davidson College in the spring semester of 2003. A proposal for green chemistry curriculum development was initiated to the Associated Colleges of the South (a consortium of 15 small liberal arts colleges in the south) in the fall of 2002, resulting in generous funding for new course development. In January 2003 the inaugural course was offered as an elective to junior and senior chemistry majors and minors, and it filled up quickly to a maximum enrollment of 16 students. The primary classroom format was lecture, but this was supplemented by visiting speakers (David Blauch, John Frost, Mary Kirchhoff, and Robin Rogers), student seminars, the ACS video presentation, a field trip, and a poster fair. This course had no laboratory component. Students were graded on two 25-minute oral seminars, one written mid-term examination, a poster presentation, and attendance/class participation. The course met on a Monday/Wednesday/Friday schedule, which allowed for class periods of 50-minutes. Often it seemed that the 50-minute class periods were limiting; perhaps a Tuesday/Thursday class schedule allowing 75-minute instructional periods would have been better. The course began using the classic text by Anastas and Warner *(2)*, which was completed in the first two weeks. The class then embarked upon Matlack's encyclopedic text *(4)* and used it for the remainder of the semester. The course syllabus is available on the web *(24)*.

Highlights of the course were seeing reticent students come out of their shell during their oral seminar and poster presentations, seeing students get genuinely excited about green chemistry, and taking a field trip to a green dry cleaner that employs a liquid carbon dioxide process in place of traditional perchloroethylene. Student interest and motivation remained high throughout the course, culminating in some of the most creative and original entries seen in years at the annual Davidson College Science Poster Fair. On their final course evaluations, some students wrote that their mode of thinking about chemistry had been forever transformed by this course, while others expressed the desire to pursue some additional type of green chemistry experience. As a direct or indirect result of this course, three students pursued green chemical industrial internships during the following two summers (at Boehringer-Ingelheim, Pfizer, and Merck), one student was accepted into the ACS Green Chemistry Summer School, and two other students went on to graduate school in environmental science and in green chemistry (at Columbia University and The Queen's University of Belfast, respectively).

Designing and teaching this new course generated a lot of positive public relations for both the institution and the instructor. Student "word of mouth" publicity spread across campus, resulting in other (future) potential students e-mailing the instructor requesting that the course to be taught more frequently. (Regrettably, due to scheduling issues, it can be offered only every other year.) Several times during the semester, short articles appeared in campus-related publications such as the student newspaper, the quarterly *Davidson College Journal*, the local newspaper, and so on. The course also was acknowledged in some of the ACS green chemistry educational materials *(18)*, as well as in a National Academy of Sciences brochure *(25)*. Certainly the institution takes note of and enjoys these sorts of public relations events. How to measure or quantify the amount of professional advancement due solely to this type of curriculum development is challenging, but the impact for all involved (institution, instructor, and students) has been very favorable.

What this course accomplished at Davidson is not unique. Other "stand-alone" lecture courses in green chemistry have been developed by several people. These include, but are not limited to, courses developed by Penny Brothers (University of Auckland), Al Matlack (University of Delaware), Terry Collins (Carnegie Mellon University), Richard Gurney (Simmons College), and Denyce Wicht (Suffolk University). At Davidson College and at other institutions, the development of novel green chemistry courses and the greening of traditional chemistry courses have been viewed as innovative by departments and administrators.

Greening Traditional Chemistry Courses

Many of us struggle with the question, "How do I go about greening my course?" One of the best things about green chemistry is that it finds applicability to practically every chemistry course, regardless of traditional discipline. For example, in an <u>organic</u> chemistry course, one might begin by emphasizing one or more of the following green themes: substitution of less toxic reagents for more toxic reagents, improvements in atom economy, reducing solvent volume (or eliminating solvent altogether), alternative sources of energy (*e.g.*, sonication, microwaves), selection of a green solvent, and/or elimination of protecting groups. (For a detailed discussion of green organic chemistry, see Chapter 3 in this volume written by Tom Goodwin.) In an <u>inorganic</u> chemistry course, catalysis (design, synthesis, recycling, turn-over numbers, *etc.*), reagents on inorganic solid supports (silica gel, alumina), solid acids and bases, silver or gold nanoparticles, and/or reduction of heavy metals can be introduced. For a biochemistry course, one could teach the chemistry of biodiesel, biocatalysis, biosynthesis, fermentation, and/or raw materials derived from renewable resources. In an analytical chemistry course, possible topics might include in-line and on-line process analytical technologies (in-process infrared spectroscopy, for example), efficient separations that avoid column chromatography or energy-intensive distillations, and/or electrochemical synthesis. (For a detailed discussion of green analytical chemistry, see chapter 7 in this volume written by Liz Gron.) In physical chemistry, one may explore the properties of carbon dioxide in each of its three physical states, the thermochemistry of biodiesel, photochemistry, kinetics and catalysis, and/or the advantages of performing computational studies. In a polymer chemistry course, minimization of wastes in polymer synthesis, the chemistry of biodegradable polymers, and/or reagents on polymeric supports could be discussed. Other ideas and examples are given in Mike Cann's excellent publication *(26)* and web site *(27)* dealing with "Greening Across the Curriculum".

Greening the Mainstream Chemistry Textbooks

Although some progress has been made in the past couple of years, by and large the greening of mainstream American chemistry textbooks remains to be addressed. As of this writing, there are three mainstream American chemistry textbooks that have substantially incorporated sections on green chemistry:

- Hill and Kolb, *Chemistry for Changing Times, 11th Edition* (28);
- Solomons and Fryhle, *Organic Chemistry, 8th Edition* (29);
- Baird and Cann, *Environmental Chemistry, 3rd Edition* (30).

Given the large number of new and revised American chemistry textbooks published annually, this list of three texts is underwhelming. Considering the bestselling mainstream American organic chemistry textbooks on the market for the "full-year" course which include Bruice *(31)*, Carey *(32)*, Eğe *(33)*, Fox and Whitesell *(34)*, Hornback *(35)*, Loudon *(36)*, McMurry (standard edition) *(37)*, McMurry (biological approach edition) *(38)*, Smith *(39)*, Solomons and Fryhle *(29)*, Vollhardt and Schore *(40)*, and Wade *(41)*, the question becomes "Why has only one set of authors chosen (so far) to green new editions of their text?" In order to get the message out to larger groups of students, this needs to change. Certainly the other authors can no longer claim ignorance of the importance of green chemistry to the future of planetary health through the reduction of hazardous substances and minimization of wastes. Students throughout the world should have access to green chemistry information within their texts. The textbook is a very powerful medium for transmission of information and, as such, needs to be designed with a focus on working to solve the problems of modern society. For a more detailed discussion of the topic of infusing green chemistry into mainstream American chemistry textbooks and working with authors and publishers for change, please see chapter 6 in this volume written by Mike Cann.

Scholarship

Scholarly activity is typically defined by the priorities of the individual institution. Some institutions may accept pedagogical development work as scholarly activity within the parameters of tenure and/or promotion portfolios. Other institutions may only accept results from the more traditionally defined research programs for similar decisions. The practice of green chemistry allows scientists at all types of institutions to further their careers in ways that enhance the mission of their own institutions. The focus of this section is on scholarly work in the development of green educational materials and on the connection between the development of these materials and the professional enhancement of faculty researchers.

While the questionnaire that was discussed in the introduction was helpful in seeing trends and generating some statistical information, it is important to back up the data with examples of scholarly activity that have been published in various formats. A literature search produced seventy-one papers related to green chemistry education materials development *(26, 42-111)*. This search was accomplished using standard chemical literature search engines and can be considered complete through June 2007. These papers spanned all fields of chemistry and included both lecture and laboratory curriculum development. Other resources, such as the Greener Education Materials for Chemists (GEMs) database *(22)* and meeting symposia abstracts *(112)*, reflect increased interest in green chemistry education materials development. Additionally, a text for non-majors has been published recently with end-of-chapter green chemistry exercises *(28)*.

Many innovative examples of laboratory curriculum development spanning the range of green chemistry principles have been published recently in the *Journal of Chemical Education*. Some examples are listed below; but due to space limitations, all of the references cannot be discussed. Grigoriy Sereda, at the University of South Dakota, in a green alternative to aluminum chloride-catalyzed alkylation of xylene, utilizes graphite to promote alkylation of *p*-xylene *(43)*. The authors claim to utilize four of the twelve green chemistry principles: waste prevention, atom economy, catalysis, and safety. Other recently published examples include solvent-free Wittig reactions *(46, 82)* and a solvent-free Baeyer-Villiger lactonization *(66)*. Water was used as a solvent in organic synthesis in the preparation of *meso*-diethyl-2-2'-dipyrromethane *(51)*.

Two preparations of biodiesel for organic laboratories *(44, 59)* have been published. An interesting example of a green enantioselective aldol condensation was published in 2006 by George Bennett at Millikin University *(50)*. This paper discusses the green principles that the experiment follows, but also addresses why it fails to be a perfect example of a green reaction. This is an issue that many struggle with when teaching or developing experiments: the gray area that makes something greener, but not perfect. Students are often locked in an "all or nothing" mindset, making it vital to provide them with the tools that teach the critical thinking skills necessary for making decisions about an experimental method.

Many schools, even though they may have a research component to tenure and promotion, have limited resources available for faculty to actually do research. This, coupled with sometimes heavy teaching loads, can make it difficult to have a continuous, productive research program. A way around some of these issues has been for faculty to utilize laboratory time to develop new curriculum. Richard Gurney from Simmons College has used laboratory time in his organic course to develop a green microwave oxidation of borneol to camphor *(113)*. This has taken place as an iterative process over the course of several semesters, with the result being a new experimental procedure that is considerably greener than the one originally in the laboratory curriculum.

Other groups have been working to develop green materials for use in lecture courses. In addition to the texts listed in the previous section, there are several recent literature examples *(45, 47, 68, 75)* that incorporate green chemistry principles with industrial and social components. Two people, Margaret Kerr and Mary Kirchhoff, have been awarded Fulbright fellowships to promote green chemistry internationally. Margaret Kerr will travel to Thailand on her sabbatical to work with faculty at Chulalongkorn University in Bangkok in developing green chemistry curriculum under the Fulbright Senior Scholars program. Mary Kirchhoff traveled to Uruguay for two weeks in November, 2006 under the Fulbright Senior Specialist program. In both cases, green chemistry was an excellent "big picture" way to address some globally recognized problems.

The GEMs database *(22)*, while duplicating some already published works, also provides educators with the opportunity to post and use education materials prior to publication. Some materials development may not be extensive enough to publish in a peer-reviewed journal. In order to have those materials available to the general public, however, the educator is able to post single experiments or course syllabi. As this type of publication contributes to the overall knowledge base of the green chemistry community, smaller institutions with less of a research focus consider this to be part of a scholarly portfolio.

An example of this is an experiment developed by undergraduate researchers under the direction of Margaret Kerr at Worcester State College. The organic chemistry laboratory was using a bromination of alkenes experiment from the Doxsee and Hutchison laboratory manual *(15)*. This particular experiment, using an *in situ* generation of Br_2 from HBr and H_2O_2 to brominate *trans*-stilbene, is very effective in illustrating green chemistry principles while demonstrating standard organic laboratory techniques. Faculty teaching the general-organic-biochemistry course for nursing students were interested in incorporating this experiment into their curriculum. The experiment, however, was not designed for non-majors and would be too advanced for students with no other chemistry experience. Work done by undergraduate researchers over a semester produced an experiment that is suitable for non-majors or for use as a demonstration. Cyclohexene is used in place of *trans*-stilbene due to its solubility and overall higher rate of reaction. The experiment is done at room temperature without the use of any solvent and incorporates only a round-bottomed flask and a condenser as a safety measure to prevent the escape of any bromine vapor. The reaction is very fast, with the color of the bromine disappearing within five minutes. It is highly effective as a visual demonstration of halogenation of alkenes. This reaction has been successfully implemented in the general-organic-biochemistry course curriculum.

Another experiment uses two known green reactions to synthesize 2,3-dibromo-1,3-diphenyl-1-propanone, as shown in Figure 1. The first step is the formation of a chalcone *via* an aldol condensation of benzaldehyde and acetophenone that was developed by Daniel Palleros at the University of California, Santa Cruz *(84)*. This step is solvent free and can be accomplished in a test tube. The second step is a bromination of the alkene of the chalcone using *in situ* formation of Br_2 from HBr and H_2O_2 as described above. The second step is accomplished in a round-bottomed flask with a condenser attached to prevent escape of any Br_2 vapor. The chalcone is dissolved in a minimal amount of warm ethanol and then added to the Br_2 dropwise. The reaction is over within 5-8 minutes and forms the final product in >90% yield as an easily isolated white powder. Other published syntheses of this molecule utilize less green reagents, although there is some work being done to use alternative brominating agents *(114-119)*. Similar to the experiment above, this set of reactions can be easily used in a non-majors lab to demonstrate carbon-carbon bond formation *via* an aldol condensation followed by halogenation of an alkene. There are no complicated glassware requirements and the final product is easily isolated. This set of reactions also can be utilized in the standard organic chemistry curriculum by incorporating techniques such as recrystallization, melting points, and spectroscopic analyses. This experiment can be completed in one lab period and demonstrates several green principles: solvent and waste minimization, atom economy, and the use of safer reagents.

While the two experiments described above may be considered to be derivative in a peer-reviewed publication, they are an important addition to the toolbox of green reactions available for a course for non-majors and for the expansion of green reactions for the organic laboratory. By utilizing the GEMs

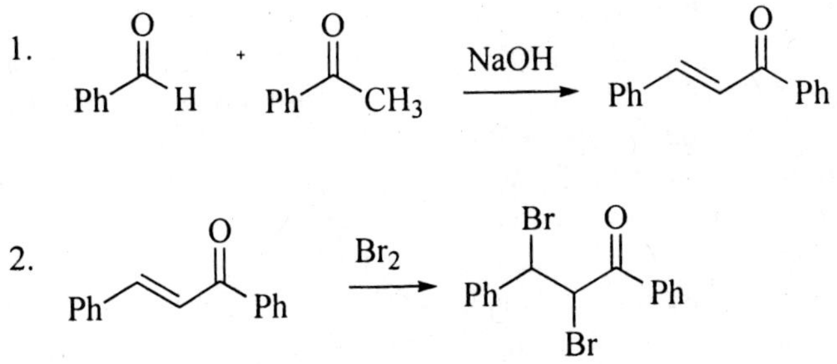

(Br_2 formed *in situ* via the reaction of $HBr + H_2O_2$)

Figure 1. Formation of 2,3-dibromo-1,3-diphenyl-1-propanone.

database submission process, this work has been made available for others who may require a similar experiment. It was presented by students as a poster at a conference on undergraduate research. The research was considered professional development, as it enhanced the overall teaching quality of the department and fostered the undergraduate research program.

Service

The third component to many tenure and promotion decisions is often referred to as "service". This has many definitions and is generally split into service for the college and service to the outside community. Green chemistry is being utilized as a way for chemists to reach out to the community in a positive way and to involve students in service learning type projects. Expansion of green chemistry into the K-12 curriculum has been fostered by outreach programs developed by many people.

Nearly fifty percent of respondents to the previously mentioned questionnaire indicated that they utilize green chemistry in outreach programs. The type of outreach that is being accomplished is as varied as the faculty members who are doing it. Some examples are discussed below.

Worcester State College has a diverse student body, reflecting its urban population. More than sixty languages are spoken in the public schools within the city. Typically in a class at Worcester State College, there are several students who do not speak English as their first language. In order to utilize the language skills of such a diverse student body, a translation project was started. Two students translated and tested the green bromination of cyclohexene, discussed in the previous section, into Vietnamese. The experiment will be available for use in high schools in the city and also in the international outreach programs through the ACS Green Chemistry Institute. Another program at Worcester State College provides science demonstrations to Spanish speaking students in 3rd through 6th grades at an after school program focusing on tutoring and an introduction to scientific principles. Bilingual students were employed to act as translators for those faculty members who do not speak Spanish. Faculty from biology, computer science, physics and chemistry all provided demonstrations and activities throughout the year. Topics such as sustainability, the environment, and introduction to chemistry were introduced during the sessions. This program has just completed its first year and is expected to continue for subsequent years. The goal is to have students attend for multiple years so that demonstrations and ideas can become more complex, with the use of green chemistry as a foundation for building scientific literacy.

Other faculty members have utilized the American Chemical Society student affiliate chapters for green chemistry outreach. Many of these student affiliate chapters have been awarded "green" status through the ACS GCI. In order to be awarded this status, the chapter must complete at least three activities that are considered to be green in nature *(19)*. The chapter must submit a report documenting their results. Thirty-seven chapters have been awarded green status since the inception of the program in 2002 *(120)*. Some chapters—such as Bridgewater State College and University of Tennessee, Martin—have documented their activities through presentations at ACS national meetings. These programs have led a variety of activities including green magic shows, high-school outreach, inviting speakers to campus, presentations about biodiesel and the importance of soybeans, and designing t-shirts with the green principles on them.

Other examples of outreach were given by respondents to the questionnaire, many of whom recognize the importance of including green chemistry into K-12 science education. These outreach activities included demonstrations and classroom visits, seminars and workshops for teachers, science camps, Earth Day activities, ACS project SEED programs, college tours, and National Chemistry Week activities. Green chemistry faculty have organized and participated in various activities involving high school teachers throughout the nation.

Conclusion

Herein we have highlighted numerous contributions from faculty members who have enhanced their professional growth and development by using the creation of green chemistry educational materials as the vehicle. Green chemistry has served—and will continue to serve for the foreseeable future—as a fruitful discipline for activities that enhance one's professional development, whether in teaching, scholarship, or community and/or institutional service. The future looks bright and we welcome you to join us in this beneficial, exciting, and rewarding endeavor.

Acknowledgements

M. E. K. would like to acknowledge the members of the Worcester State College chemistry department and administration for their support and encouragement in the development of a green chemistry program. Financial support from Worcester State College, University of Oregon and the Green Chemistry Institute is gratefully acknowledged. She would also like to thank student researchers Bankole Samuel, Emmanuel Jones, Angela Gikunju, Aaron Stairs, Khanh Vo, Heidi Lam and Jessica Sargent for their hard work and spirited discussions.

D. M. B. would like to thank the administration and chemistry faculty of Davidson College for their support and encouragement during the development and teaching of CHE 304. Financial support for curriculum development from the Associated Colleges of the South is gratefully acknowledged. Also acknowledged are many valued contributions from his talented research students: Richard Clary, Daniel Cooke, Jamie Causey, Charles Lea, and Isaac Miller.

Both authors would like to acknowledge Ms. Kay Filar of Davidson College for editorial assistance.

References

1. Leitner, W. *Green Chem.* **2004**, *6*, 351.
2. Anastas, P. T.; Warner, J. C. *Green Chemistry: Theory and Practice*; Oxford University Press: Oxford, U.K., 1998.
3. Ritter, S. K. *Chem. Eng. News* **2007**, *85(22)*, 38-40; also see: http://www.gcednet.org/ (accessed Feb 19, 2008).
4. Matlack, A. S. *Introduction to Green Chemistry*; Marcel Dekker, Inc.: New York, 2001.
5. Lancaster, M. *Green Chemistry: An Introductory Text*; Royal Society of Chemistry: Cambridge, U.K., 2002.
6. Manahan, S. E. *Green Chemistry and the Ten Commandments of Sustainability*, 2nd ed.; ChemChar Research, Inc.: Columbia, MO, 2005.
7. Stevens, E. S. *Green Plastics: An Introduction to the New Science of Biodegradable Plastics*; Princeton University Press: Princeton, NJ, 2001.
8. Clark, J. H.; Macquarrie, D. *Handbook of Green Chemistry and Technology*; Blackwell Publishing Limited: Oxford, U.K., 2002.
9. *Ionic Liquids as Green Solvents: Progress and Prospects*; Rogers, R. D., Seddon, K. R., Eds.; American Chemical Society: Washington, DC, 2003.

10. Koichi, M. *Green Reaction Media in Organic Synthesis*; Blackwell Publishing Limited: Oxford, U.K., 2005.
11. *Chemistry for Green Environment*; Srivastava, M. M., Sanghi, R., Eds.; Alpha Science International, Ltd.: Oxford, U.K, 2005.
12. *Ionic Liquids: Industrial Applications to Green Chemistry.* American Chemical Society: Washington, DC, 2006.
13. Sheldon, R. A.; Arends, I.; Hanefeld, U. In *Green Chemistry and Catalysis*; Rogers, R. D., Seddon, K. R., Eds.; Wiley-VCH: New York, 2007.
14. *Greener Approaches to Undergraduate Chemistry Experiments*; Kirchhoff, M., Ryan, M. A., Eds.; American Chemical Society: Washington, DC, 2002.
15. Doxsee, K. M.; Hutchison, J. E. *Green Organic Chemistry: Strategies, Tools, and Laboratory Experiments*; Thomson Brooks/Cole: Belmont, CA, 2004.
16. Cann, M. C.; Connelly, M E. *Real-World Cases in Green Chemistry*; American Chemical Society: Washington, DC, 2000.
17. *Introduction to Green Chemistry: Instructional Activities for Introductory Chemistry*; Ryan, M. A., Tinnesand, M., Eds.; American Chemical Society: Washington, DC, 2002.
18. *Going Green: Integrating Green Chemistry into the Curriculum*; Parent, K., Kirchhoff, M., Godby, S., Eds.; American Chemical Society: Washington, DC, 2005.
19. ACS Green Chemistry Institute Home Page. http://www.greenchemistry institute.org/ (accessed Feb 19, 2008).
20. U.S. Environmental Protection Agency. Green Chemistry Home Page. http://www.epa.gov/greenchemistry/ (accessed Feb 19, 2008).
21. Green Chemistry Network Home Page. http://www.rsc.org/Chemsoc/ GCN/index.htm (accessed Feb 19, 2008).
22. The Greener Education Materials (GEMs) for Chemists Database. http://greenchem.uoregon.edu/gems.html (accessed Feb 19, 2008).
23. Weise, E. *Green Chemistry: Innovations for a Cleaner World* [VHS]; American Chemical Society: Washington, DC, 2000.
24. Brown, D. M. CHE 304 Syllabus. http://www.chm.davidson.edu/courses/ 304Bsyl.pdf (accessed Feb 19, 2008).
25. National Research Council of the National Academies. *Sustainability in the Chemical Industry: Grand Challenges and Research Needs*; The National Academy Press: Washington, DC, 2006; pp 104-105.
26. Cann, M. C.; Dickneider, T. A. *J. Chem. Educ.* **2004**, *81*, 977-980.
27. Greening Across the Chemistry Curriculum Home Page. http://academic.scranton.edu/faculty/CANNM1/dreyfusmodules.html (accessed Feb 19, 2008).
28. Hill, J. W.; Kolb, D. K. *Chemistry for Changing Times*, 11th ed.; Pearson Prentice Hall: Upper Saddle River, NJ, 2007.

29. Solomons, T. W. G.; Fryhle, C. B. *Organic Chemistry*, 8th ed.; Wiley: New York, 2003.
30. Baird, C.; Cann, M. *Environmental Chemistry*, 3rd ed.; W. H. Freeman and Co.: New York, 2004.
31. Bruice, P. Y. *Organic Chemistry*, 5th ed.; Pearson Prentice Hall: Upper Saddle River, NJ, 2006.
32. Carey, F. A. *Organic Chemistry*, 6th ed.; McGraw-Hill: Boston, 2005.
33. Eğe, S. *Organic Chemistry: Structure and Reactivity*, 5th ed.; Houghton Mifflin Company: Boston, 2003.
34. Fox, M. A.; Whitesell, J. K. *Organic Chemistry*, 3rd ed.; Jones and Bartlett Publishers: Sudbury, MA, 2004.
35. Hornback, J. M. *Organic Chemistry*, 2nd ed.; Thomson Brooks/Cole: Belmont, CA, 2005.
36. Loudon, G. M. *Organic Chemistry*, 4th ed.; Oxford University Press: New York, 2001.
37. McMurry, J. *Organic Chemistry*, 6th ed.; Thomson Brooks/Cole: Belmont, CA, 2003.
38. McMurry, J. *Organic Chemistry: A Biological Approach*; Thomson Brooks/Cole: Belmont, CA, 2007.
39. Smith, J. G. *Organic Chemistry*, 2nd ed.; McGraw-Hill: Boston, 2007.
40. Vollhardt, K. P. C.; Schore, N. E. *Organic Chemistry*, 5th ed.; W. H. Freeman and Co.: New York, 2006.
41. Wade, L. G. *Organic Chemistry*, 6th ed.; Pearson Prentice Hall: Upper Saddle River, NJ, 2006.
42. Roberts, K. *Educ. Chem.* **2007**, *44*, 43-44.
43. Sereda, G. A.; Rajpara, V. B. *J. Chem. Educ.* **2007**, *84*, 692-693.
44. Bucholtz, E. C. *J. Chem. Educ.* **2007**, *84*, 296-298.
45. Marteel-Parrish, A. E. *J. Chem. Educ.* **2007**, *84*, 245-247.
46. Nguyen, K. C.; Weizman, H. *J. Chem. Educ.* **2007**, *84*, 119-120.
47. Heaton, A.; Hodgson, S.; Overton, T.; Powell, R. *Chem. Educ. Res. Pract.* **2006**, *7*, 280-287.
48. Richardson, A.; Janiec, A.; Chan, B. C.; Crouch, R. D. *Chem. Educator* **2006**, *11*, 331-333.
49. Veiga, M. A. M. S.; Rocha, F. R. P. *Chem. Educator* **2006**, *11*, 187-189.
50. Bennett, G. D. *J. Chem. Educ.* **2006**, *83*, 1871-1872.
51. Sobral, A. J. F. N. *J. Chem. Educ.* **2006**, *83*, 1665-1666.
52. Pereira, J.; Afonso, C. A. M. *J. Chem. Educ.* **2006**, *83*, 1333-1335.
53. Bennett, J.; Meldi, K.; Kimmell, C., II. *J. Chem. Educ.* **2006**, *83*, 1221-1225.
54. Braun, B.; Charney, R.; Clarens, A.; Farrugia, J.; Kitchens, C.; Lisowski, C.; Naistat, D.; O'Neil, A. *J. Chem. Educ.* **2006**, *83*, 1126-1129.
55. Ravia, S.; Gamenara, D.; Schapiro, V.; Bellomo, A.; Adum, J.; Seoane, G.; Gonzalez, D. *J. Chem. Educ.* **2006**, *83*, 1049-1051.
56. Cacciatore, K. L.; Sevian, H. *J. Chem. Educ.* **2006**, *83*, 1039-1041.

57. Mak, K. K. W.; Siu, J.; Lai, Y. M ; Chan, P. K. *J. Chem. Educ.* **2006**, *83*, 943-946.
58. Touchette, K. M. *J. Chem. Educ.* **2006**, *83*, 929-930.
59. Clarke, N. R.; Casey, J. P.; Oneyma, E.; Donaghy, K. J. *J. Chem. Educ.* **2006**, *83*, 257-259.
60. Iacobucci, S.; Jaworek-Lopes, C.; Wang, P.; Phun, L.; Wilbur, G.; Dobson, A. *Chem. Educator* **2006**, *11*, 102-104.
61. Musiol, R.; Tyman-Szram, B.; Polanski, J. *J. Chem. Educ.* **2006**, *83*, 632-633.
62. Montes, I.; Sanabria, D.; Garcia, M.; Castro, J.; Fajardo, J. *J. Chem. Educ.* **2006**, *83*, 628-631.
63. Dintzner, M. R.; Wucka, P. R.; Lyons, T. W. *J. Chem. Educ.* **2006**, *83*, 270-272.
64. Crumbie, R. L. *J. Chem. Educ.* **2006**, *83*, 268-269.
65. Akers, S. M.; Conkle, J. L.; Thomas, S. N.; Rider, K. B. *J. Chem. Educ.* **2006**, *83*, 260-262.
66. Esteb, J. J.; Hohman, J. N.; Schlamadinger, D. E.; Wilson, A. M. *J. Chem. Educ.* **2005**, *82*, 1837-1838.
67. Boatman, E. M.; Lisensky, G. C.; Nordell, K. J. *J. Chem. Educ.* **2005**, *82*, 1697-1699.
68. Van Arnum, S. D. *J. Chem. Educ.* **2005**, *82*, 1689-1692.
69. Cheung, L. L. W.; Lin, R. J.; McIntee, J. W.; Dicks, A. P. *Chem. Educator* **2005**, *10*, 300-302.
70. Bennett, G. D. *J. Chem. Educ.* **2005**, *82*, 1380-1381.
71. Romero, A.; Hernández, G.; Suárez, M. F. *J. Chem. Educ.* **2005**, *82*, 1234-1236.
72. White, L. L.; Kittredge, K. W. *J. Chem. Educ.* **2005**, *82*, 1055-1056.
73. Jones-Wilson, T. M.; Burtch, E. A. *J. Chem. Educ.* **2005**, *82*, 616-617.
74. Soltzberg, L. J.; Brown, V. *J. Chem. Educ.* **2005**, *82*, 526.
75. Grant, S.; Freer, A. A.; Winfield, J. M.; Gray, C.; Lennon, D. *Green Chem.* **2005**, *7*, 121-128.
76. Cave, G. W. V.; Raston, C. L. *J. Chem. Educ.* **2005**, *82*, 468-469.
77. McKenzie, L. C.; Huffman, L. M.; Hutchison, J. E. *J. Chem. Educ.* **2005**, *82*, 306-310.
78. Daley, J.; Landolt, R. G. *J. Chem. Educ.* **2005**, *82*, 120-121.
79. McCarthy, S. M.; Gordon-Wylie, S. W. *J. Chem. Educ.* **2005**, *82*, 116-119.
80. Correia, P. R. M.; Siloto, R. C.; Cavicchioli, A.; Oliveira, P. V.; Rocha, F. R. P. *Chem. Educator* **2004**, *9*, 242-246.
81. Esteb, J. J.; Gligorich, K. M.; O'Reilly, S. A.; Richter, J. M. *J. Chem. Educ.* **2004**, *81*, 1794-1795.
82. Leung, S. H.; Angel, S. A. *J. Chem. Educ.* **2004**, *81*, 1492-1493.
83. McKenzie, L. C.; Thompson, J. E.; Sullivan, R.; Hutchison, J. E. *Green Chem.* **2004**, *6*, 355-358.

84. Palleros, D. R. *J. Chem. Educ.* **2004**, *81*, 1345-1347.
85. Goodwin, T. E. *J. Chem. Educ.* **2004**, *81*, 1187-1190.
86. Fringuelli, F.; Piermatti, O.; Pizzo, F. *J. Chem. Educ.* **2004**, *81*, 874-876.
87. Mooney, D. *Chem. Health Saf.* **2004**, *11*, 24-28.
88. Song, Y.; Wang, Y.; Geng, Z. *J. Chem. Educ.* **2004**, *81*, 691-692.
89. McKenzie, L. C.; Huffman, L. M.; Parent, K. E.; Hutchison, J. E.; Thompson, J. E. *J. Chem. Educ.* **2004**, *81*, 545-548.
90. Seen, A. J. *J. Chem. Educ.* **2004**, *81*, 383-384.
91. Uffelman, E. S.; Doherty, J. R.; Schulze, C, Burke, A. L.; Bonnema, K. R.; Watson, T. T.; Lee, D. W., III. *J. Chem. Educ.* **2004**, *81*, 325-329.
92. Santos, E. S.; Garcia, I. C. G.; Gomez, E. F. L. *J. Chem. Educ.* **2004**, *81*, 232-238.
93. Uffelman, E. S.; Doherty, J. R.; Schulze, C; Burke, A. L.; Bonnema, K. R.; Watson, T. T.; Lee, D. W., III. *J. Chem. Educ.* **2004**, *81*, 182-185.
94. Grant, S.; Freer, A. A.; Winfield, J. M.; Gray, C.; Overton, T. L.; Lennon, D. *Green Chem.* **2004**, *6*, 25-32.
95. Wellman, W. E.; Noble, M. E. *J. Chem. Educ.* **2003**, *80*, 537-540.
96. Giokas, D. L.; Paleologos, E. K.; Karayannis, M. I. *J. Chem. Educ.* **2003**, *80*, 61-64.
97. Lennon, D.; Freer, A. A.; Winfield, J. M.; Landon, P.; Reid, N. *Green Chem.* **2002**, *4*, 181-187.
98. Harper, B. A.; Rainwater, J. C.; Birdwhistell, K.; Knight, D. A. *J. Chem. Educ.* **2002**, *79*, 729-731.
99. Pohl, N.; Clague, A.; Schwarz, K. *J. Chem. Educ.* **2002**, *79*, 727-729.
100. Warner, M. G.; Succaw, G. L.; Hutchison, J. E. *Green Chem.* **2001**, *3*, 267-270.
101. Raston, C. L.; Scott, J. L. *Pure Appl. Chem.* **2001**, *73*, 1257-1260.
102. Ware, S. A. *Pure Appl. Chem.* **2001**, *73*, 1247-1250.
103. Reed, S. M.; Hutchison, J. E. *J. Chem. Educ.* **2000**, *77*, 1627-1629.
104. Collard, D. M.; Jones, A. G.; Kriegel, R. M. *J. Chem. Educ.* **2001**, *78*, 70-72.
105. Hulce, M.; Marks, D. W. *J. Chem. Educ.* **2001**, *78*, 66-67.
106. French, L. G.; Stradling, S. S.; Dudley, M.; DeBottis, D.; Parisian, K. *Chem. Educator* **2001**, *6*, 25-27.
107. Hjeresen, D. L.; Schutt, D. L.; Boese, J. M. *J. Chem. Educ.* **2000**, *77*, 1543-1544.
108. Szafran, Z.; Singh, M. M.; Pike, R. M. *Educ. Quim.* **2000**, *11*, 172-173.
109. Singh, M. M.; Szafran, Z.; Pike, R. M. *J. Chem. Educ.* **1999**, *76*, 1684-1686.
110. Cann, M. C. *J. Chem. Educ.* **1999**, *76*, 1639-1641.
111. Collins, T. J. *J. Chem. Educ.* **1995**, *72*, 965-966.
112. For example, see: Haack, J. A.; Hutchison, J. E.; Kirchhoff, M. M.; Levy, I. J. *J. Chem. Educ.* **2005**, *82*, 974-976.

113. See: http://acs.confex.com/acs/green07/techprogram/P41698.HTM (accessed Feb 19, 2008); http://faculty.northseattle.edu/tfurutani/che238/borneol.pdf (accessed Feb 19, 2008).
114. Kavala, V.; Naik, S.; Patel, B. K. *J. Org. Chem.* **2005**, *70*, 4267-4271.
115. Lakouraj, M. M.; Tajbakhsh, M.; Mokhtary, M. *J. Chem. Res.* **2005**, 481-483.
116. Salazar, J.; Dorta, R. *Synlett* **2004**, 1318-1320.
117. Rajesh, R.; Sarma, R. G. V. S.; Sengottuvelu, S.; Vijay Kumar, S. G.; Rajan, D. S.; Suresh, B. *Indian Drugs* **2003**, *40*, 37-40.
118. Nair, V.; Panicker, S.; Augustine, A.; George, T. G.; Thomas, S.; Vairamani, M. *Tetrahedron* **2001**, *57*, 7417-7422.
119. Kabalka, G. W.; Yang, K.; Reddy, N. K.; Narayana, C. *Synth. Commun.* **1998**, *28*, 925-929.
120. Parent, K. ACS Green Chemistry Institute. Personal communication.

Figure 1.1. Growth in global green chemistry research and education. (top) In the early 1990s, green chemistry efforts were focused mainly in the United States, Italy, and the United Kingdom. (middle) The late 1990s and early 2000s saw green chemistry networks and GCI chapters make inroads in South America, Asia, and Oceania. (bottom) As of 2008, green chemistry initiatives are underway in Africa, and gains on virtually every continent have been made.

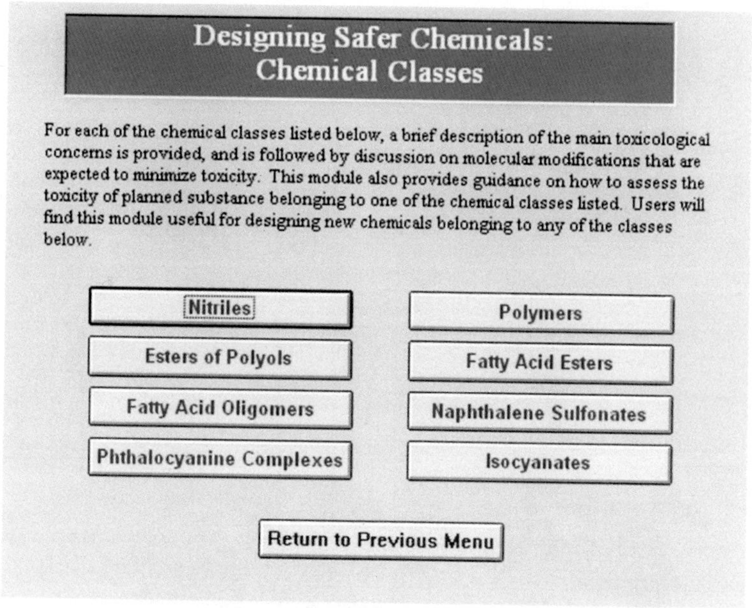

Figure 1.2. The Designing Safer Chemicals module of EPA's Green Chemistry Expert System.

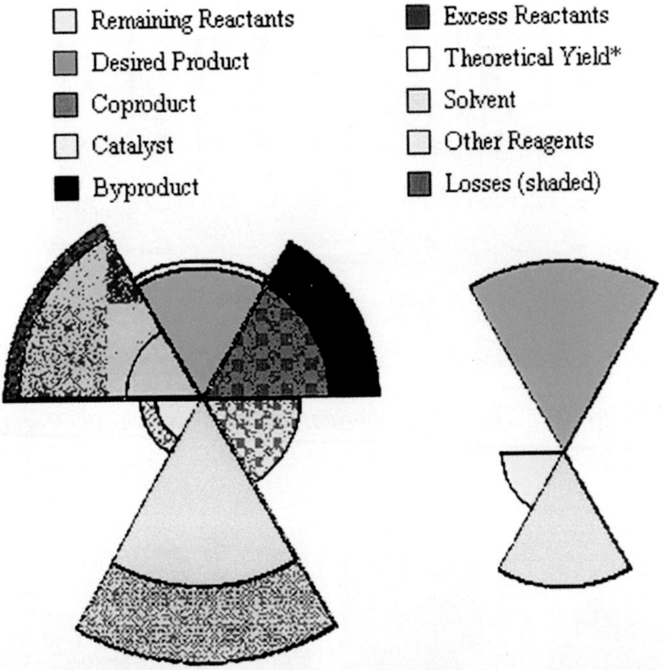

Figure 1.3. Visualizing atom economy, process efficiency, and waste generation using the Green Chemistry Assistant web application developed by St. Olaf College. The diagram on the left shows a process that uses excess reagents, generates coproducts and byproducts, has poor solvent and catalyst recovery, and involves a reaction that does not go to completion. In contrast, the diagram on the right depicts a 100% atom-economical reaction using small amounts of solvent and catalyst that are completely recovered. Such graphical representations help the user quickly identify problem areas in a chemical process.

Figure 10.1. The Green laboratory at the University of Oregon.

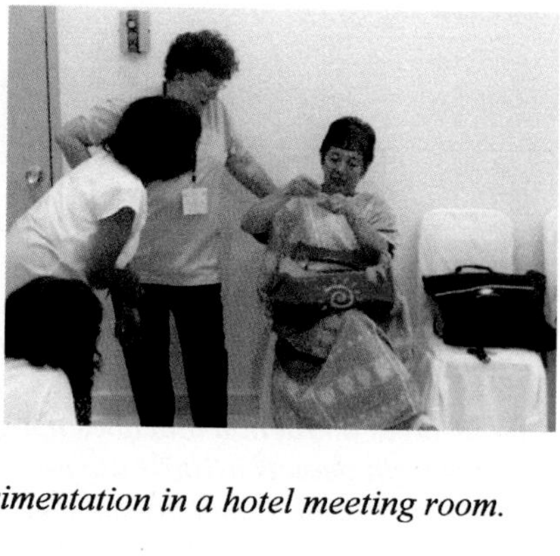

Figures 10.2 and 10.3. Organic experimentation in a hotel meeting room.

Chapter 3

The Garden of Green Organic Chemistry at Hendrix College

Thomas E. Goodwin

Department of Chemistry, Hendrix College, Conway, AR 72032

The Hendrix College organic chemistry laboratories were converted to microscale experiments in 1988 to minimize possible adverse environmental impact, increase lab safety, and decrease generation of waste and costs of waste disposal. As we became aware of the green chemistry movement in university research labs and chemical industry in 2000, we wiped the slate clean and did a thorough reevaluation of our laboratory philosophies, practices, procedures, and experiments. An account of our ruminations and conclusions has been published (*J. Chem. Educ.* **2004**, *81*, 1187-1190). This chapter, while briefly reviewing our green lab philosophies, will focus primarily on the presentation of practical green organic experiments. Some of these have been adapted from prior experiments, while others we have modified from the primary literature including our own research. We believe that as the menu of green experiments grows in size and variety, the energy of activation for going green at more colleges and universities will be lowered, to the benefit of us all.

Introduction

If a garden is aesthetically pleasing and astonishingly productive, we say that the gardener has a "green thumb". Surely a green thumb is a product of knowledge and expertise, robust seeds and seedlings, fertile soil, hard work, and cooperative weather. Furthermore, "*organic* gardening" often connotes environmentally-friendly gardening practices. But where does one learn the precepts, priorities, and practices that produce an *organic* chemist's *green* thumb in the laboratory? There is likely no one-size-fits-all recipe for success, and no program is perfect, but a study of current practitioners of the art is a good place to start. A garden of green organic chemistry is growing at Hendrix, and therein lies a tale. In the telling we will follow the instructions of the King to the White Rabbit in *Alice in Wonderland* (*1*): "Begin at the beginning and go on till you come to the end, then stop."

We converted almost completely to microscale organic chemistry experiments around 1988, primarily for green reasons. It was our belief, which has been born out in practice, that we could successfully teach basic organic chemistry lab practices, principles and thought processes using smaller amounts of reagents than those that had been used traditionally. One tenet of green chemistry is to reduce environmental waste; that is certainly guaranteed if one downsizes reagent quantities. Of course microscale does not automatically mean green, but the combination of greener and smaller scale experiments would seem to be a win-win situation (*2*). Although making our experiments more environmentally-friendly had long been our objective, we soon became more conscious of the green chemistry movement through journal and magazine articles. Aided by a donation from an alumnus, we developed some green organic experiments, and presented our initial work at the ACS National Meeting in San Diego in Spring 2001. At that meeting, we learned of the upcoming Green Chemistry in Education workshop at the University of Oregon in July 2001, organized by Ken Doxsee, Jim Hutchison, and their colleagues. The author attended that first workshop as a participant, and then was invited back for several subsequent iterations as a speaker. These were highly educational and inspirational meetings, which were superb networking opportunities.

As we approached the deliberate, comprehensive greening of our organic chemistry laboratories, we began at the outset to think through our reasons for having labs in the first place. This was followed by an evaluation of the experiments that we were doing or considering to articulate what each was intended to teach, and how they could be greened or replaced with more environmentally-friendly alternatives. We quickly realized the following: (1) merely making an experiment greener than a previous version does not necessarily make it environmentally benign; (2) no experiment is perfectly risk-free; (3) the best that we can do is to reevaluate continually and try to make experiments better. As Anastas and Warner stated, "...the goal of making a chemical product or process 'environmentally benign' is a mere statement of the ethic of continuous improvement more than it is a metric by which to measure improvement (*3*)." We have previously described our "asymptotic approach" to the development of a green organic chemistry laboratory in more detail (*4*). We often modify and adapt existing experiments, but also develop new ones, with a special focus on carbon-carbon bond formation. Our objectives, thought processes, and accomplishments are best illustrated by examples. Therefore, a few snapshots will be presented below to show our version of the variety of experiments that can be introduced into the green organic laboratory. We'll start with a few simple experiments adopted from prior work, and employed to introduce basic lab techniques and equipment. This will be followed by a discussion of some new and more in-depth experiments that we have developed at Hendrix. We believe that as the number of available green chemistry experiments increases, the energy barrier toward a move to green labs will be greatly lowered.

The Early Experiments

We strive to coordinate our laboratory experiments with what we are covering at the time in the classroom. The early part of most organic chemistry classes consists of laying the foundation: a review of salient parts of General Chemistry, learning nomenclature, studying stereochemistry, being introduced to the arcana and jargon of the subdiscipline, and so on, with few chemical reactions included. Therefore, our early experiments focus on introducing lab techniques, glassware, and procedures, with spectroscopic analysis being added as it is covered in class. One of our objectives is to assist our students with better risk assessment skills in the evaluation of chemicals and chemical processes. Reading Material Safety Data Sheets (MSDS) is important, but these alone do not always give one a good sense of perspective. Caffeine, for example, is stated to be "toxic if swallowed; irritating to eyes, respiratory system, and skin; target organs: central nervous system, heart". Nonetheless, we know from our personal experiences that drinking beverages containing caffeine poses no serious health risk.

We find that many common lab techniques and procedures can be taught easily using commercially available preparations or natural products. For example, our students are introduced to microscale distillation using a Hickman distillation head to distill acetone or ethyl acetate from commercial nail polish remover. We also often use natural products as a starting point to introduce lab techniques. Although it is certainly not a novel thought to teach in the organic lab via isolation of natural products, for there is a long history of doing so, these can also be used as a starting point for green experiments. Three classic procedures which we have adopted and adapted from others will suffice to illustrate these points.

Biosynthesis of Ethanol from Molasses

Production of ethanol via yeast-catalyzed fermentation of plant carbohydrates is an ancient process. Professor John Thompson (Lane Community College, Eugene, Oregon, USA) has developed an interesting variant of this process using molasses as the feedstock. We use this experiment to introduce the ideas of catalysis (yeast enzymes), azeotropes, density, and biofuels, as well as the technique of simple distillation of ethanol using 19/22 glassware (*5*).

Isolation and Saponification of Trimyristin: A Fat from Nutmeg

This experiment was first published in 1971 using diethyl ether to extract the fat trimyristin from the common spice, nutmeg (*6a*). It has since appeared in many organic chemistry lab texts, and was expanded to include soap preparation by saponification of the trimyristin (*6b*). [This recent incarnation elicited an exchange of letters regarding the hazards of distilling diethyl ether solutions in the lab (*6c)*.] The version of this experiment that we use was adapted from a remarkably prescient lab text generated at Brown University in 1974. This in-house text was entitled "Zero Effluent Laboratory", and was notable in that the practitioners made an attempt to account for and recover all wastes, by-products and solvents from each experiment, that is, to practice green chemistry (*7*). They used a type of Freon to extract trimyristin from nutmeg, as that solvent was widely deemed at the time to be environmentally benign. (*Nota bene*: this illustrates a valuable lesson; namely, the need for us always to be cautious when handling even "environmentally benign" chemicals, for today's "green chemicals" may at some point in the future be found to be less innocuous than we think at present.) We made the minor change of using petroleum ether as the extraction solvent, since it readily dissolves the relatively non-polar fat, and does not pose the environmental hazards of Freon, or the peroxide risks of diethyl ether. Students are introduced to the techniques and concepts of solvent extraction from an edible spice, filtration over a Buchner funnel, simple distillation in 19/22 glassware, recrystallization, saponification, and micelles. This is also a good opportunity to discuss DeSimone's use of liquid carbon dioxide and unique surfactants to provide a greener dry cleaning procedure that does not rely on the traditional solvent tetrachloroethylene (*8*).

Isolation of Chlorophyll and Carotenoid Pigments from Spinach

For an introduction to thin layer and gravity column chromatography, we separate the pigments from frozen spinach. This is one green experiment that is literally green. A number of variations of this classic procedure have been reported (*9*). We have made only minor changes in the procedure of Pavia and co-authors (*9d*). However, since we obtain the UV/Vis spectrum of the carotene isolated from the spinach, implementation of this experiment facilitates a discussion of ultraviolet/visible spectroscopy.

Introductory Experiments Involving Organic Reactions

As we progress further into the first semester of an introductory course in organic chemistry, specific chemical reactions are introduced. Below we will briefly discuss four experiments that exemplify green chemistry principles and facilitate teaching important concepts and procedures.

Hydrolysis of 2-Chloro-2-methylbutane, a Microscale S_N1 Reaction

Gilbert and Martin have described an excellent kinetics experiment which illustrates solvolysis of 2-chloro-2-methylbutane via an S_N1 mechanism (*10*). The reaction is carried out in a green solvent mixture of water and 2-propanol. In addition, we have scaled down to the use of only 118 microliters of 2-chloro-2-methylbutane per student, and carry out our titrations using reusable plastic syringes instead of burets (*11*). This lab exercise provides experimental support for the proposed S_N1 mechanism and illustrates the effect of solvent polarity (increasing percentages of water in the solvent mix) on reaction rate.

Friedel-Crafts Alkylation of 1,4-Dimethoxybenzene: No $AlCl_3$

The Friedel-Crafts alkylation is a classic illustration of the general class of electrophilic aromatic substitutions. Traditionally, Friedel-Crafts reactions require an alkyl halide as the electrophile source and at least a molar equivalent of aluminum chloride, a hygroscopic and caustic powder that is rather problematic to use in the introductory lab. An easy variant of this procedure utilizes the reactive substrate 1,4-dimethoxybenzene, *tert*-butyl alcohol as the electrophile precursor, and sulfuric acid as the catalyst (*12*). We run this reaction on a microscale, and use commercial "rubbing alcohol" (70% aqueous 2-propanol) in place of methanol as the recystallization solvent. The product, 1,4-di-*tert*-butyl-2,5-dimethoxybenzene, exhibits simple ^1H and ^{13}C NMR, and mass spectra, thus providing a good introduction to these spectroscopic analyses. We add value to the experiment and introduce molecular modeling by having students use Spartan (Wavefunction, Inc.) to minimize the structures of the mono-, di-, tri-, and tetra-*tert*-butylated products. This helps to visualize the increase in steric strain that explains why only the symmetrical di-*tert*-butyl product is produced. It is of interest to note that with four *tert*-butyl substituents, Spartan shows a non-planar and hence non-aromatic ring. Molecular modeling is, of course, the ultimate green chemistry, squandering only some electrons.

Green Metrics: Atom Economy is Not the Whole Story

We often have students assess their reactions using various green metrics, of which atom economy is the most familiar (*13*). To calculate the percent atom economy starting from the balanced reaction equation, one divides the molar mass of the desired product by the molar masses of the reactants and multiplies by 100 (*14*). For this Friedel-Crafts reaction, the percent atom economy is a relatively high 87.4%. However, atom economy does not take into account the actual percent yield isolated by the student. This calculation has been called the percent experimental atom economy (*15*), and may be illustrated in the following way. If we suppose our student researcher got a percent yield of 60%, then multiplying 0.60 by 87.4% we get 52.44%; not quite so impressive. These calculations, however, don't tell the whole story. In this reaction one uses 120 mg of 1,4-dimethoxybenzene, 525 mg of glacial acetic acid, 155 mg of *tert*-butyl alcohol, and 920 mg of concentrated sulfuric acid, as well as approximately 1.3 g of 70% 2-propanol as a recrystallization solvent. Hudlicky and co-workers described a green metric which they called the "effective mass yield", in which the isolated mass of the desired product is divided by the sum of the masses of non-benign materials in the reaction (*16*). For the purposes of illustration, let us assume that all of the materials in our Friedel-Crafts reaction are non-benign, although one might argue effectively that, for example, acetic acid and 70% 2-propanol are benign. Furthermore, we will assume a 100% yield of 1,4-di-*tert*-butyl-2,5-dimethoxybenzene (217.4 mg). In this case, our effective mass yield in percent would be 7.23 %; quite unimpressive indeed, and quite an eye-opener when compared to the atom economy of 87.4%. So is this a green experiment or not? That question is a good catalyst for class discussion.

Ultramicroscale Reduction of 2,6-Dimethylcyclohexanones with Sodium Borohydride

This clever experiment was introduced by Garner and is an excellent introduction to stereochemical concepts, chair conformations of cyclohexane rings, and reduction of ketones using $NaBH_4$ (*17*). Garner carried out the reduction on one microliter ("ultramicroscale") of a mixture of *cis*- and *trans*-2,6-dimethylcyclohexanones in methanol, followed by gas chromatography. We and others have shown that the product ratio is dependent on the alcohol chosen as a solvent (*18*). We have also shown that the reduction proceeds readily using water as the only reaction solvent, thus greening the experiment even further. We add value to this experiment by Spartan molecular modeling, as well as Arrhenius calculations to compare stabilities of the multiple stereoisomers produced in the reaction. The Karplus relationship between dihedral angles and ^{1}H NMR coupling constants can also be illustrated.

Chemistry and Flatulence: An Introductory Enzyme Experiment that is Green Indeed

This experiment with an eye-catching name is quite environmentally friendly, using *green* split peas as the substrate, water as the solvent, and the commercial product Beano® as the reagent. We have altered it very little from the procedure reported by Hardee and co-workers (*19*). The experiment provides a good introduction to a discussion of carbohydrates via the use of enzymes in the Beano to catalyze hydrolysis of acetal linkages in oligosaccharides found in split green peas. The increasing concentration of glucose with time is measured using a glucometer (such as used by diabetics), and data are graphed with Microsoft Excel®.

New Green Chemistry Experiments Developed At Hendrix College

Solventless, Room-Temperature Diels-Alder Reaction & Intramolecular Nucleophilic Acyl Substitution: A Huge Increase in Molecular Complexity in a 10-mL Beaker

As an experiment for the introductory organic chemistry laboratory, McDaniel and Weekly reported the Diels-Alder reaction of (*E,E*)-2,4-hexadien-1-ol (**1**) with the popular dienophile maleic anhydride (**2**) in refluxing toluene to provide, after subsequent intramolecular nucleophilic acyl substitution, carboxylic acid **3** and its enantiomer *(20)*. We wanted to eliminate the toluene, but were unsuccessful in finding a greener solvent that worked well. Subsequently, we found success by merely heating the two reagents with no solvent at 90 °C for 15 minutes, or by microwave heating at reduced power for 30 seconds. Finally, we were surprised and delighted to discover that the reaction can be run with no solvent, simply by stirring the two reactants together at room temperature (Figure 1). We carry out the reaction on a very small scale

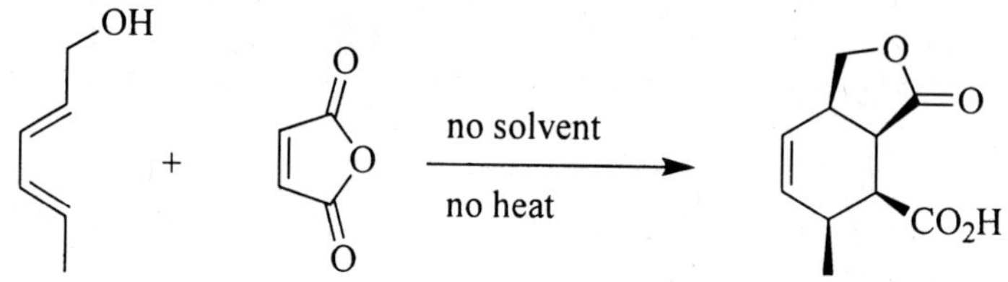

1 2 3

Figure 1. Solventless, room temperature Diels-Alder reaction.

in a 10 mL beaker. The only energy required is provided by 10-15 minutes of manual stirring by the experimentalist, for whom the ultimate energy source is the sun. The atom economy for this reaction is 100%. This simple evolution of a reaction procedure illustrates the surprising progress one can make when thinking green. While this is not an experiment in which students learn much lab technique, it is one in which they can learn a lot about green chemistry thought processes, as well as experiment selection and planning. We should always discuss with our students how and why the process evolved, as well as why we choose some experiments and reject others. As a part of each laboratory write-up, we also ask that three questions be answered: (1) What was green about this experiment?; (2) What was not green?; (3) How could we make it greener?

It is interesting to note that in this process, two achiral reactants provide only two stereoisomers out of 16 possible for a product with four chirality centers. As illustrated by McDaniel and Weekly, the product provides a good opportunity to introduce analysis of structure using 2D COSY NMR spectroscopy. We also include a Spartan modeling exercise for measuring dihedral angles and relating them to observed coupling constants, including those engendered by diastereotopic hydrogens.

Green Epoxidation of Geraniol Using 3% Hydrogen Peroxide

The conversion of alkenes to epoxides is covered in most introductory organic chemistry texts, often exemplified by the use of *m*-chloroperoxybenzoic acid (MCPBA) as the epoxidizing reagent. Bradley *et al.* compared the enantioselective Sharpless epoxidation of geraniol with the classical MCPBA method, which gives a racemic product (*21*). Hoye and Jeffrey have illustrated a "mini-microscale" version of this MCPBA reaction using only 1 mg of geraniol (*22*). Epoxidation with MCPBA exhibits a very poor atom economy, since only one oxygen atom of MCPBA ends up in the desired product. A greener epoxidation of geraniol has been reported using 30% hydrogen peroxide as the oxygen source (with a by-product of water) and a catalytic amount of methyltrioxorhenium (MTO), but is carried out in a rather undesirable solvent, dichloromethane (*23*). We used this procedure in our synthesis of some new sesquiterpenes (farnesol derivatives) that we identified in the temporal gland secretions of African elephants (*24*). We have greatly modified this procedure to make it more environmentally friendly, and have used it to develop an epoxidation of the natural product geraniol (**4**) for the introductory organic chemistry lab.

Interestingly, the MTO/H_2O_2 epoxidation of geraniol occurs at the double bond distal to the hydroxyl group to give epoxide **5**, and is thus complementary to MCPBA epoxidation at the proximal double bond. We have substantially greened up the original MTO-catalyzed epoxidation of geraniol in the following ways: (1) use of 2:1 water/95% ethanol as the solvent; (2) substitution of nicotinamide (niacinamide; from a commercial vitamin tablet) in place of the usual base (pyridine) as a catalyst activator and buffer; (3) use of over-the-counter 3% hydrogen peroxide as the oxidizing reagent (Figure 2). The desired product (or even a product mixture) lends itself well to analysis via ^{1}H NMR spectroscopy (*22*).

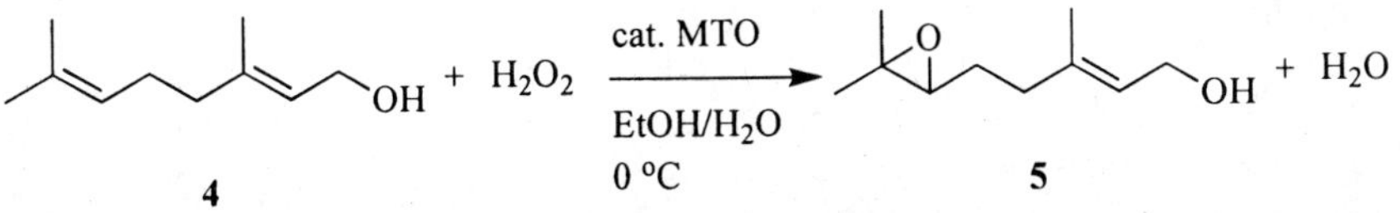

4 **5**

Figure 2. Geraniol epoxidation with hydrogen peroxide.

Solventless, Microwave-Assisted Synthesis of Coumarins via a Tandem Knoevenagel Condensation and an Intramolecular Nucleophilic Acyl Substitution

We have been particularly enamored with the development of experiments involving carbon-carbon bond formation, especially as part of tandem reactions occurring in a single container (see the Diels-Alder reaction, Figure 1). One such reaction is the synthesis of simple esters of coumarin-3-carboxylic acids via a Knoevenagel condensation between malonic esters and various α-hydroxybenzaldehydes, followed by intramolecular nucleophilic acyl substitution. This conversion, catalyzed by piperidine, has been carried out under a variety of conditions, for example, at room temperature without solvent by grinding in a mortar and pestle, microwave heating of a solventless mix of reactants, and stirring a neat mixture of reagents at the ambient temperature *(25)*. In order to introduce our students to the concept of microwave-enhanced reactions as a more energy-efficient alternative to conventional heating, we have developed our own variation of the solventless Bogdal conditions (*25b*). We react 0.5 mmol of **6** (X = H or CH_3O) and **7** with 0.05 mmol of piperidine in a watchglass-covered 10 mL beaker using a commercial microwave oven at 50% power for 2 minutes to produce coumarins **8** (Figure 3).

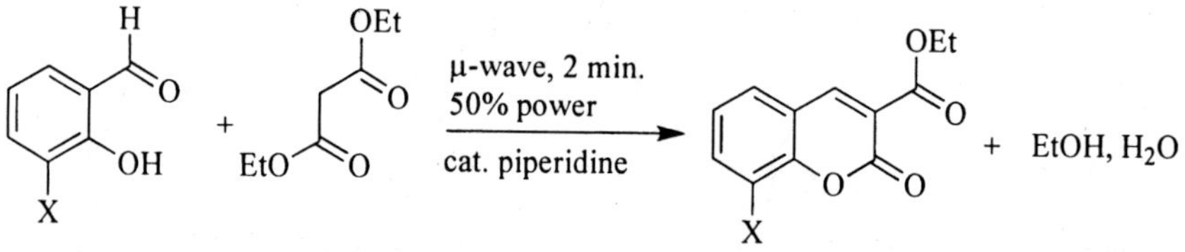

6 **7** **8**

Figure 3. Microwave-assisted synthesis of coumarins.

Palladium-Catalyzed Cross-Coupling Reactions

Palladium-catalyzed cross-couplings are among the most important reactions in modern organic chemistry for carbon-carbon bond formation (*26*). Yet, these reactions are rarely mentioned in introductory organic chemistry texts or in laboratory manuals (*27*). Below we present microscale, green variations of two such reactions that we have developed for the introductory organic chemistry laboratory.

A Greener Sonogashira Coupling

The Sonogashira reaction involves coupling between an aromatic iodide (**9**) or bromide and a terminal alkyne (**10**) to provide an aryl alkyne (**11**) *(28)*. In 1999 we reported a green, microscale variation which involved refluxing for 30 minutes in 95% ethanol, open to the atmosphere, with no added solubilizing ligand for the palladium, and using an environmentally friendly base (Figure 4) *(29)*.

These conditions are greener than those traditionally employed (*e.g.*, triphenylphosphine as the ligand, triethylamine as base, and acetonitrile as solvent). In addition to the introduction of a modern chemical transformation to the lab, the experiment can be extended by conversion of the initial coupling

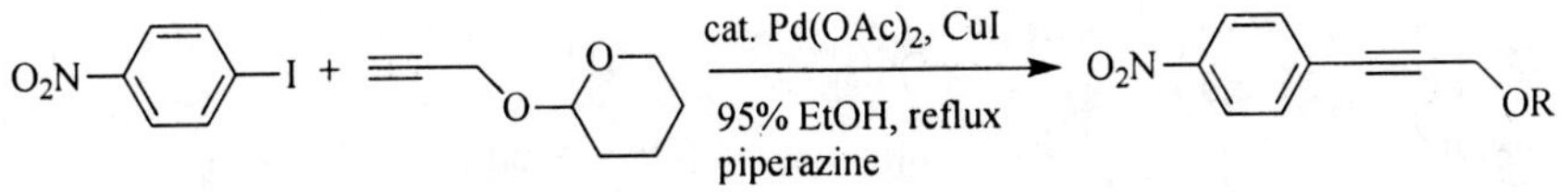

9 10 11

Figure 4. Green Sonogashira coupling.

product into two subsequent derivatives. The experiment also teaches NMR spectral analysis (including the Karplus relationship between dihedral angle and coupling constants), molecular modeling, conformational analysis, and the anomeric effect.

An Extraordinarily Green Suzuki-Miyaura Cross-Coupling

The Suzuki-Miyaura reaction is currently one of the most important carbon-carbon bond forming reactions in modern synthetic organic chemistry, and is particularly useful for forming biaryl compounds via the coupling of an aryl boronic acid and an aryl bromide or iodide *(30)*. A typical Suzuki-Miyaura procedure would include a palladium source (*e.g.*, $Pd(OAc)_2$, approximately 5 mol %) and a solubilizing ligand (*e.g.*, Ph_3P), a solvent (*e.g.*, toluene, dimethylformamide), and a base (*e.g.*, aqueous Na_2CO_3), at an elevated temperature *(31)*. In recent years, a number of more environmentally friendly procedures have been developed, including reactions in aqueous media, often with microwave heating *(32)* and in polyethylene glycol (PEG) *(33)*. Liu, Zhang, and Wang reported efficient Suzuki-Miyaura cross-couplings in mixtures of PEG and water *(34)*. The Leadbeater group has demonstrated that biaryl formation via Suzuki-Miyaura cross-coupling can occur in aqueous ethanol under microwave heating for 20 minutes using incredibly low concentrations of palladium (0.0045 mol %) *(35)*. These authors state that using conventional heating for their reaction mixtures requires hours to produce much lower reaction yields. We have blended the Zhang and Leadbeater protocols to produce an extraordinarily green and convenient procedure for the introductory organic chemistry laboratory (Figure 5).

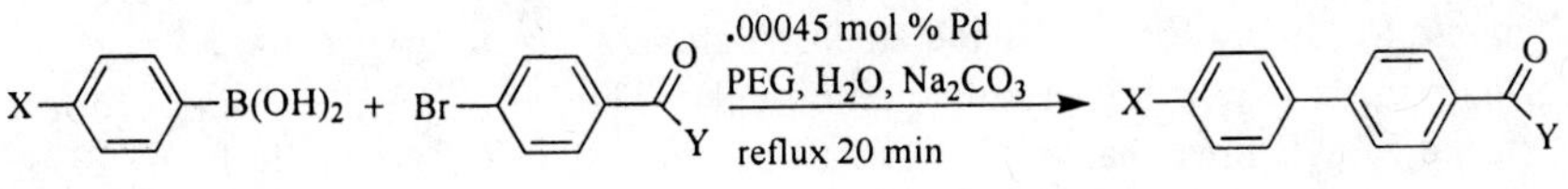

12 13 14

Figure 5. Aqueous Suzuki-Miyaura coupling.

Essentially, we stir and reflux a 1:1 molar ratio of aryl boronic acid **12** (X = CH_3 or CH_3O) and aryl bromide **13** (Y = H or CH_3) for 20 minutes in a 1:5 w/w mixture of PEG-400 and dilute aqueous Na_2CO_3 containing 0.00045 mol % Pd^0 to produce typically 70-80% yields of the desired biaryl **14** as a white solid precipitate of sufficient purity to provide good melting points and NMR spectra. The reaction does not require an inert atmosphere. PEGs are non-toxic, inexpensive, and water soluble, and are available in a wide variety of molecular weights *(36)*. We are currently exploring the range and combinations of aryl boronic acids, aryl halides, and PEGs for which these environmentally friendly reaction conditions are successful. Details of this procedure and our findings will be published in due course.

Solventless Extraction Techniques For The Study Of Volatile Organic Compounds Present In Natural Products

Traditionally, extraction of dissolved organic chemicals from dilute aqueous media was carried out via an organic solvent and a separatory funnel or an apparatus for continuous extraction. More recently, greener solvent-free headspace sampling techniques have been developed. These offer several advantages over older methodologies, including the following: pre-GC concentration of a relatively large quantity of analytes onto a very small area, ease of use, ease of automation, and faster extraction times. The first to be developed was solid phase microextraction (SPME) *(37)*. SPME (Supelco, Inc.) relies on a small glass fiber coated with an adsorbant polymer that can be exposed to the headspace over an aqueous sample, or immersed directly in the aqueous solution. Due in part to its ease of use, SPME has been employed for an immense variety of applications. SPME is easily automated using the versatile, rugged, and immensely useful Combi PAL GC autosampler (CTC Analytics). Manual SPME/GC-MS has been used in several experiments for the undergraduate laboratory *(38)*. The newest of the solventless techniques is solid phase dynamic extraction (SPDE; Chromsys, Inc.). SPDE features concentration of headspace analytes by repetitive dynamic flow back and forth over a polymer coating on the inside wall of a stainless steel syringe needle attached to a gas-tight syringe *(39)*. SPDE has more adsorbant polymer coating than SPME and is also easily automated by use of the Combi PAL robot. Unlike SPME, SPDE is a dynamic technique for headspace analysis, and appears to offer some advantages for extraction of volatile organic compounds. The SPDE needle is more robust than the SPME fiber, has more extraction capacity, and can usually be used for hundreds of extractions before replacement. SPDE/GC-MS is useful in a variety of applications *(40)*.

We have employed automated headspace SPME/GC-MS and SPDE/GC-MS in our search for elephant pheromones (*41*). Currently, we are developing a new automated SPDE/GC-MS experiment for the undergraduate lab, in which we analyze the volatile organic compounds from various types of tea *(42)*. After the analysis, the tea samples can be discarded in the trash can as they are clearly environmentally benign. We expect that the powerful, solventless, green analytical techniques of automated SPME and SPDE, coupled with GC-MS, will increasingly be used in undergraduate teaching and research labs as more faculty become familiar with them, and additional appropriate experiments are developed.

Some Final Musings

Microscale, Green Organic Chemistry

Since the inception of microscale experiments, some have argued that the use of microscale glassware and techniques does not adequately prepare a student for research in graduate school or chemical industry after graduation. We respectfully disagree, and offer the following comments for consideration. (1) Our objective should be education, not "training". For example, if we teach the principles behind extraction, or distillation, the particular application of those principles to a new type of glassware should be easy to master. (2) At Hendrix, the majority of our organic chemistry students will not pursue careers as chemists, therefore the particular size and shape of glassware utilized in the organic lab will likely not affect positively or negatively their chosen career path. (3) For those students who do plan to become chemists, they will encounter a variety of glassware in subsequent chemistry laboratories and in undergraduate research projects. In their graduate work or in chemical industry, they may run reactions in batch reactors or in 96 well plates, but well educated and resourceful students will learn fast and adapt quickly. (4) Finally, the proof is in the pudding; we have seen many students who are products of the microscale organic lab at our institution and at many others, go on to excel in top-notch graduate programs and in a variety of chemical professions.

Safe Chemistry and Green Chemistry

As we pursue green chemistry ideals, we should talk to our students about risks, and risk perception. All activities in life involve some risk. If a student drives to the nearest town in her car, she realizes at some level that there is a chance she'll have a wreck, since she knows that such wrecks do happen. Yet, she usually decides that the benefits are worth the risks. Analogously, no lab experiment that we ask a student to carry out is perfectly risk-free. But with the driving, the student makes the choice; for the lab experiment however, the instructor makes the decision. We must take that responsibility seriously. It is important to discuss possible risks with our students, and to explain how we have attempted to minimize those risks (*4*).

It may seem that we have been alluding to what has traditionally been called "lab safety", but that is only partially correct. Perhaps we should consider for a moment what we mean when we say "green chemistry" and when we say "safe chemistry". The former has been defined as "the utilization of a set of principles that reduces or eliminates the use or generation of hazardous substances in the design, manufacture, and application of chemical products", and is further delineated through the Twelve Principles of Green Chemistry (*43*). "Safe chemistry" perhaps connotes also such things as lab safety glasses and fume hoods. That is, we may be tempted to think of "green chemistry" in terms of protecting the global environment and sustainability issues, and "safe chemistry" as protecting the students in the lab. Of course, it should not be either/or, but rather both/and. Paracelsus is often credited with expressing the sentiment in the 16th century that "the dose makes the poison". This phrase, which is the title of a good toxicology text for the non-specialist, is relevant to safe, green chemistry (*44*). That is, if we do not implement prudent and generally accepted lab safety practices, or if we are mistaken about the greenness of an experiment, or if an accident happens, the student who is standing in front of the experiment will be most adversely affected and will get the largest "dose". We believe that the combination of microscale quantities, standard lab safety practices, and a commitment to the principles and procedures of green chemistry offers the best hope for protection of both the lab environment and the global environment. Many good chemists have demonstrated that we can be safer and greener, with no sacrifice in pedagogical quality.

Conclusion

Our green organic chemistry experiments and philosophies have been highlighted above, but we are greening all aspects of our chemistry laboratory program at Hendrix. Particularly notable are our introductory laboratories in General Chemistry, spearheaded by Liz Gron, and ably assisted by our colleagues Warfield Teague, David Hales, and Randy Kopper. They have developed an analytically-focused laboratory program which features environmentally friendly analysis of environmental samples. For more details see chapter 7 in this volume written by Liz Gron. This model is being extended to our lab for non-science majors, directed by Teague. Yes, a garden of green chemistry is growing at Hendrix College. We hope that our readers will be inspired to try some of the organic chemistry experiments described above, and will use their own "green thumbs" to produce some home-grown, environmentally-friendly experiments to share with others. May their organic chemistry gardens prosper, and may the green force be with them.

Acknowledgments

We are indebted to Shelly Bradley and Linda Desrochers for their dedicated assistance and advice, to our colleagues for their support and wise counsel, and to our many excellent organic chemistry students who helped troubleshoot the experiments discussed herein. We also are grateful for an early donation from an anonymous Hendrix alumnus to help us initiate our green organic chemistry and general chemistry laboratory programs.

References

1. Carroll, L. *Alice in Wonderland*; Clarkson N. Potter: New York, 1973; p 104.
2. Singh, M. M.; Szafran, Z.; Pike, R. M. *J. Chem. Educ.* **1999**, *76*, 1684-1686.
3. Anastas, P. T.; Warner, J. C. *Green Chemistry: Theory and Practice*; Oxford University Press: New York, 1998; p 15.
4. Goodwin, T. E. *J. Chem. Educ.* **2004**, *81*, 1187-1190.
5. Also see: (a) Oliver, W. R.; Kempton, R. J.; Conner, H. A. *J. Chem. Educ.* **1982**, *59*, 49-52. (b) Maslowsky, E., Jr. *J. Chem. Educ.* **1983**, *60*, 752.
6. (a) Frank, F.; Roberts, T.; Snell, J.; Yates, C.; Collins, J. *J. Chem. Educ.* **1971**, *48*, 255-256. (b) de Mattos, M. C.; Nicodem, D. E. *J. Chem. Educ.* **2002**, *79*, 94-95. Also see: (c) Umland, J. B. *J. Chem. Educ.* **2002**, *79*, 1070. (d) de Mattos, M. C.; Nicodem, D. E. *J. Chem. Educ.* **2002**, *79*, 1070. (e) Jones, J. C. *J. Chem. Educ.* **2004**, *81*, 193.
7. Corwin, L. R.; Roth, R. J.; Morton, T. H. *Zero Effluent Laboratory*, Brown University; http://www.hendrix.edu/greenchemistry (accessed Feb 26, 2008).
8. (a) McClain, J. B.; Betts, D. E.; Canelas, D. A.; Samulski, E. T.; DeSimone, J. M.; Londono, J. D.; Cochran, H. D.; Wignall, G. D.; Chillura-Martino, D.; Triolo, R. *Science* **1996**, *274*, 2049-2052. (b) Cann, M. C.; Connelly, M. E. *Real-World Cases in Green Chemistry*; American Chemical Society: Washington, DC, 2000; pp 13-18.
9. For example, see: (a) Collins, C. *J. Chem. Educ.* **1963**, *40*, 32-33. (b) Cousins, K. R.; Pierson, K. M. *J. Chem. Educ.* **1998**, *75*, 1268-1269. (c) Quach, H. T.; Steeper, R. L.; Griffin, G. W. *J. Chem. Educ.* **2004**, *81*, 385-387. (d) Pavia, D. L.; Lampman, G. M.; Kriz, G. S.; Engel, R. G. *Introduction to Organic Laboratory Techniques: A Microscale Approach*, 3rd ed.; Saunders College Publishing: New York, 1990; pp 158-164.
10. Gilbert, J. C.; Martin, S. F. *Experimental Organic Chemistry: A Miniscale and Microscale Approach*, 3rd ed.; Harcourt College Publishers: Orlando, FL, 2002; pp 442-451.

11. Postma, J. M.; Roberts, J. L., Jr.; Hollenberg, J. L. *Chemistry in the Laboratory*, 5th ed.; Freeman: New York, 2000; pp 16-17.
12. For example, see: (a) Williamson, K. *Macroscale and Microscale Organic Experiments*, 3rd ed.; Houghton Mifflin: Boston, MA, 1999; pp 366-370. (b) Mohrig, J. R.; Hammond, C. N.; Morrill, T. C.; Neckers, D. C. *Experimental Organic Chemistry*; Freeman: New York; 1998; pp 232-236.
13. For an excellent comparison of green metrics for a reaction, see: McKenzie, L. C.; Huffman, L. M.; Hutchison, J. E. *J. Chem. Educ.* **2005**, *82*, 306-310.
14. Trost, B. M. *Science* **1991**, *254*, 1471-1477.
15. See pp 5-12 of ref 8b.
16. Hudlicky, T.; Frey, D. A.; Koroniak, L.; Claeboe, C. D.; Brammer, L. E., Jr. *Green Chem.* **1999**, *1*, 57-59.
17. Garner, C. M. *J. Chem. Educ.* **1993**, *70*, A310-A311.
18. Also see: (a) Goodwin, T. E.; Meacham, J. M.; Smith, M. E. *Can. J. Chem.* **1998**, *76*, 1308-1311. (b) Hathaway, B. A. *J. Chem. Educ.* **1998**, *75*, 1623-1624.
19. (a) Hardee, J. R.; Montgomery, T. M.; Jones, W. H. *J. Chem. Educ.* **2000**, *77*, 498-500. Also see: (b) Wang, J.; Macca, C. *J. Chem. Educ.* **1996**, *73*, 797-800.
20. McDaniel, K. F.; Weekly, R. M. *J. Chem. Educ.* **1997**, *74*, 1465-1467.
21. Bradley, L. M.; Springer; J. W.; Delate, G.M.; Goodman, A. *J. Chem. Educ.* **1997**, *74*, 1336-1338.
22. Hoye, T. R.; Jeffrey, C. S. *J. Chem. Educ.* **2006**, *83*, 919-920.
23. (a) Villa de P., A. L.; De Vos, D. E.; Montes de C., C.; Jacobs, P. A. *Tetrahedron Lett.* **1998**, *39*, 8521-8524; (b) See also, Kuhn, F. E.; Herrmann, W. A. *Chemtracts* **2001**, *14*, 59-83.
24. Goodwin, T. E.; Brown, F. D.; Counts, R. W.; Dowdy, N. C.; Fraley, P. L.; Hughes, R. A.; Liu, D. Z.; Mashburn, C. D.; Rankin, J. D.; Roberson, R. S.; Wooley, K. D.; Rasmussen, E. L.; Riddle, S. W.; Riddle, H. S.; Schulz, S. *J. Nat. Prod.* **2002**, *65*, 1319-1322.
25. (a) Sugino, T.; Tanaka, K. *Chem. Lett.* **2001**, *30*, 110-111. (b) Bogdal, D. *J. Chem. Res. (S)* **1998**, *8*, 468-469. (c) Aktoudianakis, E.; Dicks, A. P. *J. Chem. Educ.* **2006**, *83*, 287-289.
26. For example, see: (a) Tsuji, J. *Palladium Reagents and Catlalysts: Innovations in Organic Synthesis*; Wiley: New York, 1995; (b) Li, J. J.; Gribble, G. W. *Palladium in Heterocyclic Chemistry*; Tetrahedron Organic Chemistry Series, vol. 20, Elsevier Science Ltd: Oxford, 2000)
27. For notable exceptions, see: (a) Clayden, J.; Greeves, N.; Warren, S.; Wothers, P. *Organic Chemistry*; Oxford University Press: Oxford, 2001; (b) Doxsee, K. M.; Hutchison, J. E. *Green Organic Chemistry, Strategies, Tools, and Laboratory Experiments*; Thomson Brooks/Cole: Belmont, CA, 2004.

28. Sonogashira, K.; Tohda, Y.; Hagihara, N. *Tetrahedron Lett.* **1975**, *19*, 4467-4470.
29. Goodwin, T. E.; Hurst, E. M.; Ross, A. S. *J. Chem. Educ.* **1999**, *76*, 74-75.
30. Miyaura, N.; Suzuki, A. *Chem. Rev.* **1995**, *95*, 2457-2483.
31. For example, see: (a) Handy, S. T.; Zhang, Y.; Bregman, H. *J. Org. Chem.* **2004**, *69*, 2362-2366. (b) Callam, C. S.; Lowary, T. L. *J. Chem. Educ.* **2001**, *78*, 947-948.
32. (a) Franzen, R.; Xu, Y. *Can. J. Chem.* **2005**, *83*, 266-272. (b) Leadbeater, N. E. *Chem. Commun.* **2005**, 2881-2902.
33. Namboodiri, V. V.; Varma, R. S. *Green Chem.* **2001**, *3*, 146-148.
34. Liu, L.; Zhang, Y.; Wang, Y. *J. Org. Chem.* **2005**, *70*, 6122-6125.
35. (a) Arvela, R. K.; Leadbeater, N. E.; Sangi, M. S.; Williams, V. A.; Granados, P.; Singer, R. D. *J. Org. Chem.* **2005**, *70*, 161-168. (b) Leadbeater, N. E.; Williams, V. A.; Barnard, T. M.; Collings, Jr., M. J. *Org. Proc. Res. Dev.* **2006**, *10*, 833-837.
36. Harris, J. M. *Poly(Ethylene Glycol) Chemistry: Biotechnical and Biomedical Applications*; Plenum Press: New York, 1992.
37. Pawliszyn, J. *Solid Phase Microextraction: Theory and Practice*; Wiley-VCH: New York, 1997.
38. (a) Pawliszyn, J.; Yang, Min J.; Orton, M. L. *J. Chem. Educ.* **1997**, *74*, 1130-1132. (b) Galipo, R. C.; Canhoto, A. J.; Walla, M. D.; Morgan, S. L. *J. Chem. Educ.* **1999**, *76*, 245-248. (c) Witter, A. E.; Klinger, D. M.; Fan, X.; Lam, M.; Mathers, D. T.; Mabury, S. A. *J. Chem. Educ.* **2002**, *79*, 1257-1260. (d) Hardee, J. R.; Long, J.; Otts, J. *J. Chem. Educ.* **2002**, *79*, 633-634.
39. Lipinski, J. *Fresenius' J. Anal. Chem.* **2001**, *369*, 57-62.
40. See, for example: (a) Musshoff, F.; Lachenmeier, D. W.; Kroener, L.; Madea, B. *J. Chromatogr. A* **2002**, *958*, 231-238. (b) Bicchi, C.; Cordero, C.; Liberto, E.; Rubiolo, P.; Sgorbini, B. *J. Chromatogr. A* **2004**, *1024*, 217-226.
41. (a) Goodwin, T. E.; Rasmussen, L. E.L.; Schulte, B. A.; Brown, P. A.; Davis, B. L.; Dill, W. M.; Dowdy, N. C.; Hicks, A. R.; Morshedi, R. G.; Mwanza, D.; Loizi, H. *Chemical Signals in Vertebrates 10* Mason, R. T.; LeMaster, M. P.; Muller-Schwarze, D. Eds.; Springer: New York, 2005; pp 128-139. (b) Goodwin, T. E.; Eggert, M. S.; House, S. J.; Weddell, M. E.; Schulte, B. A.; Rasmussen, L. E. L. *Journal of Chemical Ecology* **2006**, *32*, 1849-1853.
42. Fuller, D.; Ray, S.; Washington, C.; Broederdorf, L.; Jackson, S.; Goodwin, T., manuscript in preparation.
43. See reference 3; p 30.
44. Ottoboni, M. A. *The Dose Makes the Poison*, 2nd Ed.; Wiley: New York, 1997.

Chapter 4

Integrating Green Chemistry throughout the Undergraduate Curriculum via Civic Engagement

Richard W. Gurney and Sue P. Stafford

Simmons College, 300 The Fenway, Boston, MA 02115

The quantity of information that a chemical educator feels obligated to convey to majors and non-majors alike has resulted in the overly formulaic delivery of content that is often devoid of contextual frameworks. Correspondingly, efforts to deliver *Chemistry in Context* have begun to gain widespread popularity. Empirical evidence over the past 7 years at Northwestern University and Simmons College has indicated that introducing green chemistry both in context and through civic engagement is particularly effective. When introduced to Green Chemistry in context, students are driven to learn so that they are able to effectively educate and advocate for green chemistry in their community. Herein, three distinctly different green chemistry lecture-based courses are detailed and the design and outcome of the corresponding civic-engagement projects are described.

Our experiences over the past seven years indicate that Green Chemistry lecture based courses encourage and inspire individuals to civic action. Regardless of the audience, upon introduction to the fundamentals of chemistry through a green chemistry paradigm both majors and non-majors routinely are shocked that green chemistry is not as widespread as its merits would suggest. Tangential threads within topical, facilitated-asynchronous, online discussions appear within the first few weeks of courses, in which students attempt to understand the barriers that exist. Ultimately, students conclude that the largest barrier is in educating society, particularly their community, and they are called to action.

Herein, three distinctly different green chemistry lecture-based courses are detailed and the design and outcome of the corresponding civic-engagement projects are described. Students enrolled in a science-major, senior-seminar course organized a day-long "Undergraduate Green Chemistry Symposium" for the scientific community at Northwestern University and a week long "Green Chemistry Exhibition" at Simmons College. Students enrolled in a non-science major topical Honors Program course researched green consumer products and informed the community by submitting a weekly Green Chemistry article to the student newspaper and by creating a green version of the "What to bring to college" guide for incoming students. From these experiences we have designed and proposed what we believe will be the most effective method to educate non-science majors in green chemistry: a learning-community course coupling green chemistry and environmental ethics.

Green Chemistry as an Upper-Level, Capstone Seminar Course

The Chemistry Department of Northwestern University offered a course entitled "Green Chemistry" in the Spring 2002 and 2003 quarters and in the Honors Program at Simmons College in the Fall of 2004. The advanced level course was designed for junior or senior level undergraduate students in Chemistry, Biochemistry, Civil, Chemical and Environmental Engineering, Materials Science and related fields. The only prerequisites were general and organic chemistry. The primary goal of the upper level course was to expose students to current topics affecting the chemical industry in a multidisciplinary environment; therefore, the majority of the course material was extracted directly from the current literature (>1998). The course introduced the concept and discipline of green chemistry and placed the field's growth and expansion in a historical context from its birth in the early 1990's through the most recent Presidential Green Chemistry awards. The course introduced the 12 principles of green chemistry (*1*) as well as the tools of green chemistry including the use of alternative feedstocks or starting materials, reagents, solvents, target molecules, and catalysts. Particular attention focused on the application of innovative technology in the development of "greener" routes to improve industrial processes and to produce important green consumer products.

Asynchronous, facilitated online-discussions of current literature provided the foundation for thoughtful, content-rich discussions in the classroom. A sampling of weekly discussion prompts and their corresponding primary literature references are given in Table I. *Green Chemistry: An Introductory Text,* by Mike Lancaster, served as the primary text for the course. Course evaluation was based upon facilitated online literature discussions, literature summaries, weekly assignments, an annotated bibliography, midterm paper, and final project presentation. Sample course syllabi, weekly assignments, student papers and final presentations for all courses described herein are available from the GEMs database (*2*).

The main course project involved researching a Presidential Green Chemistry Challenge Award. Students were encouraged to refer to *Real World Cases in Green Chemistry (3)* for an example on how to structure their midterm paper. The final project for the course, originally scheduled to be a 15 minute in class presentation of the student's research, evolved into a student-organized and student-lead campus-wide Undergraduate Symposium on Green Chemistry, which included presentations from two local Presidential Green Chemistry Challenge Award Winners, Undeo-Nalco and Donlar Corporation. More than 80 members of the community were in attendance.

Student projects focused primarily on presenting the chemistry involved in various Presidential Green Chemistry Challenge Awards. However, as can be seen in Table II, a few students chose projects that could have direct application at their university.

The shift towards civic engagement for the research and the presentations was most noticeable in the topics students chose to research for the week long "Green Chemistry Exhibition" at Simmons College (bottom of Table II). Choices were driven by an upcoming building renovation and construction occurring on campus. Student topics focused on educating the community as to the upcoming choices that were to be made in the purchase of various textiles (EcoWorx™ Carpet Tile: A Cradle-to-Cradle Product, Interface Carpets, Donlar Polyaspartates, DuPont Petretec Polyester Regeneration Technology), green roofing materials (Sarnafil Inc.), dyed fabrics (Climatex Lifecycle, Rohner Textil AG), ink (soy-based), paper (Process Chlorine Free, TAML Oxidant Activators) and general knowledge of chemical safety. More than 700 student-designed pamphlets detailing the specific green chemistry projects were distributed over the course of the week. Due to the high-profile location of the Exhibition, it was estimated that more than one thousand members of the community interacted with the displays, including approximately 10 – 12 entire classes.

Table I. Weekly Topics for Upper-Level Capstone Seminar Course, Coordinated with *Green Chemistry: An Introductory Text*

Week	*Topic and Facilitated Online Discussion Prompts*
1	*Principles and Concepts of Green Chemistry* What are the general areas of investigation in green chemistry? (*4*) What is green chemistry? (*5*) What is atom economy? (*6*) How can atom efficiency be applied? (*7*)
2	*Waste: Production, Problems and Prevention* In what way(s) is click chemistry related to green chemistry? (*8*) Explain sustainable development and the triple bottom line. (*9*) How does polymer regeneration differ from standard recycling methods? (*10*) Are the properties of recycled plastics as good as the properties of virgin plastics? (*11*) Discuss the life cycle of plastics and why reducing is better than recycling and reusing. (*12*)
3	*Measuring and Controlling Environmental Performance* Why is Life Cycle Assessment important? (*13*) How can analytical techniques be used to follow the course of a reaction? (*14*) What is the Toxics Release Inventory? (*15*) What are green metrics? Are mass and energy good enough indicators of environmental impact? (*16*)
4	*Catalysis and Green Chemistry* Why are solid, unmodified alumina, silicas, and zeolites able to be used as catalysts? (*17*) Are lanthanide catalysts really environmentally friendly? (*18*) What are solid acids and solid bases and how can they be used for catalysis? (*19*) Why is catalysis a foundational pillar of green chemistry? (*20*) What are the advantages of a totally chlorine free bleaching process? (*21*)
5	*Organic Solvents: Environmentally Benign Solutions* Are fluorous solvents green? (*22*) Are ionic liquids green? (*23*) Is supercritical CO_2 a green solvent? (*24*) When can a solvent can be classified green? (*25*)
6	*Renewable Resources* What is biocatalytic synthesis? (*26*) How do fuel cells work? (*27*) How can chemicals be produced from renewable resources? (*28*) Why is the inexpensive production of levulinic acid important commercially? (*29*) What is biodiesel? (*30*)
7	*Emerging Greener Technologies and Alternative Energy Sources* What are the advantages of microwave-assisted synthesis? (*31*) How can electrochemical methods be applied to synthesis? (*32*) How can sonochemistry be applied to synthesis? How can reactions incorporate photochemical methods as an alternative energy source? (*33*) What is process intensification?
8	*Industrial Case Studies (34)*

Table II. Sample Presentation Topics

Undergraduate Green Chemistry Symposium at Northwestern University
ACQ Preserve®: A Dramatic Pollution Prevention Advancement
An Efficient Process for the Production of Cytovene®, A Potent Antiviral Agent
ULTIMER®: A Water-Soluble Polyelectrolyte for Wastewater Treatment
Manufacturing a Biodegradable Polymer from Renewable Resources: Poly(lactic acid)
MixAlco Process® of Biomass Conversion to Mixed Alcohol Fuels
Sentricon Termite Colony Elimination System®: A Chitin Synthesis Inhibitor
Conversion of Cellulosic Biomass to Levulinic Acid
THPS Biocides®: A New Class of Antimicrobial Chemistry
Designing an Environmentally Friendly Fire-Fighting Agent: Pyrocool®
Transition Metal Catalyzed Reactions for Aqueous and Air Environments
Green Chemistry at NU–The Petretech Process and Recycling of Plastics
Use of "Active" MnO_2-Silica in the 'Dry', Microwave-Assisted Oxidation of Borneol in an Undergraduate Laboratory
Green Chemistry Exhibition at Simmons College
Rethinking Carpeting: Green Design in the Carpet Industry
Traditional and Green Polymers in Your Carpet
Chemical Hazard Rating System: What You Should Know About Chemical Safety
The Science Behind a Green Roof
Ink: Know What You Print
Remaking the Way We Make Products: Improving the Paper Bleaching Process
The Environmental Impact of Paper Processing & Green Chemistry Innovations
Petroleum Dependent Development: The Unique Problem of the Petro-State

Green Chemistry as an General Education Honors Course*

A general-education course with no science prerequisites presents several additional challenges. Fundamental concepts in chemistry must be introduced but also interwoven within contextual societal frameworks to demonstrate the need for Green Chemistry. Additionally, instructors must confront student's misconceptions, fears and strong biases about the field of chemistry learned

* Students' majors in this course included: Environmental Studies, International Relations, French/International Relations, Arts Administration with a minor in French and Photography, International Relations/Economics, Political Science, Mathematics/Economics with a minor in Philosophy, Political Science/Secondary Education, Psychology, Graphic Design/Communications Media, Spanish, and International Relations/Spanish.

primarily from the environmental tragedies related to the chemical industry. What were once viewed as great advances in society from the chemical industry, evident in the Dow Chemical marketing campaign "Better things for better living though Chemistry" are now commonplace and overshadowed by more recent environmental tragedies: Bhopal, Love Canal, Chernobyl, Exxon Valdez, Cuyahoga River and Times Beach Missouri. Current traditional students matured entirely during an era in which environmental command-and-control legislation experienced a greater-than exponential increase.

Rather than ignore environmental tragedies related to the chemical industry we chose to bring them to the forefront of discussion and forge direct links to the green chemical solutions. The causal links between students' consumptive behaviors and the tragedies also provided greater student interest and involvement. A semester-long course highlighting greened products from *Real World Cases in Green Chemistry* (*3*) in the context of environmental tragedies from *Watersheds* (*37 – 44*), paced along with a traditional general, organic and biochemistry text, *Chemistry in Context* (*35*) is outlined in Table III. Alternative design paradigms for consumer products were also discussed (*36*).

Course evaluation was based upon facilitated-online literature discussions of the topics in *Watersheds* and *Real World Cases*, weekly homework assignments and quizzes from *Chemistry in Context* and a project that developed over the course of the semester. With past successful projects in mind, students designed projects with the intent of effecting a greater behavioral change in their community. Students wished to do more than simply educate the community through a one-day symposium on a Presidential Green Chemistry Challenge award winner. Students also were determined to focus their civic-engagement project such that the decision to adopt a greener consumer product would rest in the hands of students rather than administrators, as was the case in the "Green Chemistry Exposition" (Table II). Therefore, students focused their civic engagement project towards empowering students, rather than administrators, to adopt greener consumer products. The idea to create a "How to Survive College and Still Be Environmentally Friendly Guide" arose in an online discussion in which students were frustrated in attempting to locate lists of greener alternatives that are currently available:

> "if we created a list of main stream products that students purchase and use on a regular basis...then we could research...the ingredients/ chemicals, learn about those chemicals, research greener alternatives to those chemicals and processes, and then find the actual product/s that would replace the one they initially used."

During the initial project conception, students identified a resource, "the Guide" that was not readily available and championed its creation.

Multiple, higher-level learning opportunities arose as students pursued the process highlighted in the discussion post above. Students independently

Table III. A Sample Topical Schedule for a General Education Green Chemistry Course for Non-Science Majors

Chemistry in Context	*Watersheds*	*Real World Cases*
The Air We Breathe	Bhopal (*37*)	The Invention and Commercialization of a New Family of Insecticides
Protecting the Ozone Layer	Ozone (*38*)	Design and Application of Surfactants for Carbon Dioxide
The Chemistry of Global Warming	Of Greenhouses and Freezers (*39*)	The Concept of Atom Economy
Energy, Chemistry and Society	Fueling the World (*40*)	Carbon Dioxide as an Environmentally Friendly Blowing Agent
The Water We Drink	Chlorine Sunrise (*41*)	TAML Oxidant Activators: General Activation of Hydrogen Peroxide for Green Oxidation Processes
Neutralizing the Threat of Acid Rain		Designing an Environmentally Safe Marine Antifoulant
The Fires of Nuclear Fission	Chernobyl (*42*)	Production and Use of Thermal Polyaspartate Polymers
The World of Plastics and Polymers	Cradle-to-Cradle (*35*)	DuPont Petretec Polyester Regeneration Technology
Manipulating Molecules and Designing Drugs	Ingenuity of Bugs (*43*)	The BHC Company Synthesis of Ibuprofen
Genetic Engineering and the Chemistry of Heredity	Genetically Modified Organisms (*44*)	Use of Microbes as Environmentally Benign Synthetic Catalysts.

identified concepts in chemistry that they required to understand the products they were researching. Rather than being presented with a need, students' themselves found a need for the specific topical knowledge of chemistry. Lectures evolved over the course of the semester to include for example the chemistry behind creating light (incandescent, luminescent), cleaning objects (soap, detergent, pH, acidic cleaners), bleaching (oxidation), dyeing of fabrics (origin of color, mordents, structure of cotton), dyeing and curling of hair (nitrosamines, peptides, proteins, disulfide linkages), portable energy (batteries), flame retardants (halon), and plastics (polymers, plasticizers). Requests for additional lectures eventually outnumbered the remaining class time.

Students engaged in scientific method through researching their products, in the absence of a laboratory. Frustration in grandiose project design and hypothesis creation leads to constant reevaluation and new hypothesis generation. Students learned that greening is a continuous process and not a destination, for example, compact fluorescent light bulbs (CFBs) are more energy efficient, however they utilize mercury; hence, the need for reclamation centers and recycling. Also, the amount of mercury in a bulb is still less than that entering the environment from a coal-fueled power plant to run the incandescent bulb over the lifetime of the CFB (*45*). In the most successful projects, students located several subsets of problems and solutions, and began to argue their hypotheses in terms of cost-benefit analysis and life-cycle assessment.

Perhaps most striking was the paradigm shift that occurred in student's opinion of the current state of green chemical research. Initially, non-science majors nearly unanimously believed that all research in chemistry was almost complete. Initial project design was built upon the premise that a green chemical alternative existed for every non-green product or process. Student opinion changed as their research progressed and frustrations grew from their inability to locate green solutions to their product's problems. For several students, it was inconceivable that not only consumer products hazardous for human health (*e.g.*, hair dye and nitrosamines) or the environment are still on the market but also that greener alternatives have yet to be discovered. For some students this information stymied their research completely, for others it provided much needed inspiration and challenge.

What was once incontrovertible blame and contempt for the chemical industry, such as with the 'unanticipated effects' of the products chemists designed (*e.g.*, ozone depletion with CFC refrigerants, mutagenicity of thalidomide) morphed steadily into an appreciation for the nature of chemical research. Environmental tragedies began to be viewed as research opportunities through new chemistry-informed eyes. Students learned to appreciate the volume of research that still needs to be envisioned and accomplished. Students reached a new level of appreciation for the true impact of "chemistry" on their lives. Sample student projects from green chemistry as a general education course can be found in Table IV.

Tangible student outcomes included a year-long column in the student newspaper on green chemical products and a "How to Survive College and Still Be Environmentally Friendly Guide." Writing a succinct article on the chemistry/green chemistry of a consumer product to a novice challenged students to achieve a higher level of understanding of the material. Students were required to immediately engage their newfound knowledge and articulate information with their new language in a forum that would be judged not only by the professor but also their peers.

A college-wide Sustainability Committee, chartered by faculty, staff and students, was also championed and initially organized by members of the class. One of the committee's first major accomplishments was the implementation of a major sustainability initiative on campus, which was the direct result of a student project. The student extended her required exploratory research into the science and economics of compact fluorescent light bulbs (CFBs) voluntarily into the community with minimal extra encouragement. Even though the Simmons Facilities Department already switched more than 90% of lighting on campus to CFBs, the lighting in rooms within residence halls was in large measure student provided. Through a survey of a single residence hall and a few simple calculations, the student was able to prove that Simmons College could reduce not only their energy bill but also the amount of pollution the bulbs contribute to the Boston air, while providing free CFBs to all students living in the dorms. Specifically,

> "if Simmons were to purchase 317 compact fluorescent bulbs at the current catalog cost of $10.32 for Mesick resident students to use and then return at the end of the semester or academic year, it would cost $3,271.44. Based on the total cost $43,543.12 (energy cost plus bulb cost) of using incandescent bulbs compared to the total cost $11,256.67 (energy cost plus bulb cost) of using CFBs, students in Mesick can save enough money by decreasing their energy usage to overly compensate Simmons for purchasing CFBs by saving a grand total of $32,286.45 for 10,000 hours of bulb use."

Ultimately, the student's research project lead to a new Simmons policy whereby CFBs are freely distributed to all members of the community for use on campus. In the Fall of 2007, a bulb became an integral component of each student's welcome package to campus upon moving into the residence halls.

Conscience and Consumption: Green Chemistry as a Learning-Community Course with Environmental Ethics

Conscience and Consumption* is a learning community course designed to focus on the intersection of the fields of green chemistry and environmental ethics. Green chemistry involves the invention, design and application of chemical products and processes to reduce or to eliminate the use and generation of hazardous substances. Environmental ethics is the study of the nature and extent of our moral obligations to the environment. Together, these two fields of inquiry raise issues and provide answers to very practical questions concerning:

- our relation to the natural world of animals, plants, and the land,
- our use and consumption of natural resources,

* Course to be offered Fall of 2009.

Table IV. Sample Student Projects from Green Chemistry as a General Education Honors Course

Rechargeable Batteries for the Portable World
Dyes and Fabrics: Environmentally Sustainable Alternatives
The Chemical Hazards and Greener Alternatives of Dorm Furniture
Household Cleaners: Toxic Hygiene
Cool Your House, Don't Kill the Planet: Green Refrigerants
The Hidden Dangers of Polyvinylchloride
Green Manufacture of Pharmaceuticals
Toxic Computers: The Hidden Horrors of E-waste Recycling
Do Tampons Really Need to be White? Dioxins and Your Body
Paper Production and the Hidden Hazards
"Pretty" Nasty - Phthalates in Cosmetics
Simmons Dorm Cleaners and a New Green Cleaning Initiative
Cut All the Costs: Switch to Fluorescent Light Bulbs
Nitrosamines: Why I Will Never Dye My Hair Again!

- our use of technology to alter the natural world,
- our relations to the environments of other countries and cultures, and
- our obligations to advocate for change.

Conceptually, a learning community is a group of people committed to a path of inquiry and action who engage in face-to-face and online activities over an extended period of time and evolve and learn from one another. We envision that the four main dimensions of a successful learning community are comfort, trust, responsibility, and spontaneity with flexible boundaries. Logistically, the two courses are linked through an integrative seminar and the same cohort of students attends each course and the Seminar.

The six main objectives of the learning community are: (1) to provide each student with both a theoretical, scientific and practical understanding of the nature and range of environmental problems; (2) to help students develop critical thinking skills, writing skills, and laboratory skills necessary to the recognition, formulation, and assessment of environmental problems and proposed solutions; (3) to provide the opportunity for students to develop their creative thinking skills and advocacy abilities through various in-class activities and final projects; (4) to help students develop and defend their own environmental ethic; (5) to demonstrate how to formulate reasonable projects that can be accomplished in a given amount of time; and (6) to help students learn how to design, conduct, summarize and present a research project that involves social conscience.

The Seminar is an interdisciplinary arena designed to appeal to and stimulate students' increasing social conscience. By integrating the theory of moral philosophy with the practical knowledge of green chemistry, the Seminar will provide students with a unique opportunity: they will be able to approach environmental problems with a deep understanding of the assumptions and world views underlying and generating those problems, and apply scientific knowledge as a means of solving those problems. The aims of the Seminar are to provide a grounding in key issues in environmental ethics, to raise awareness of differing viewpoints, and to understand the consequences for the actions of individuals, organizations and government. The course will be driven by an activist approach, and this, coupled with group projects, will give students an opportunity to see their knowledge and values made concrete.

The Seminar is structured around two-week modules. The general learning objectives for each two-week module are to integrate knowledge of science and ethics, research the facts and background of a case area, recognize a problem or a place for improvement, see the ethical implications, research alternatives and trade-offs, evaluate alternatives scientifically and ethically and be able to recommend and defend a course of action. The specific questions that will be addressed include:

- (Weeks 2 & 3) What is the impact of our general consumption on the environment? How does green consumption lessen the impact on the environment?
- (Weeks 4 & 5) What is the impact of our oil consumption on the environment? How can we minimize our oil consumption or the impact of our oil consumption on the environment?
- (Weeks 6 & 7) What is the impact of our water use on the environment? What is the impact of our lifestyle on the environment?
- (Weeks 8 & 9) What is the impact of our energy consumption on the environment? How can we minimize our energy consumption or the impact of our energy consumption on the environment?
- (Weeks 10 & 11) How are our consumption and our use of plastics improving and destroying our lives at the same time? How can we design a new paradigm for the use and generation of plastics?
- (Weeks 12 & 13) What are we going to do about it? Be the CHANGE you wish to see in the world.

To illustrate one two-week module consider the following: the tragedy that occurred in Bhopal was created due to the release of a poisonous-gaseous chemical cloud, which was released, in large concentrations into the atmosphere. Upon reading the Bhopal case study, students are confronted with foreign terminology and several new conceptual questions. For example, one might consider the following dialogue occurring between a philosopher (P) and a chemist (C):

P: What really happened at Bhopal? What exploded?

C: A tank containing a chemical called methyisocyanate.

P: What's methylisocyanate?

C: It is a chemical used in the production of a pesticide. The molecule is a hazardous intermediate in the synthesis of a somewhat more benign chemical.

P: What's a molecule? For that matter, what is a chemical? Aren't all chemicals hazardous?

C: Oh, dear!

P: Why is it so dangerous?

C: Well, it reacts violently with water in the presence of metal to produce a toxic gas that is dispelled into the air. It quickly traveled through the air and people's eyes and particularly their lungs were severely damaged.

P: Only people? Didn't it kill animals too? Aren't their deaths part of the disaster?

C: Well I guess, methylisocyanate would have a nearly identical impact on animals. But animals aren't people so their lives aren't as important, are they?

P: They feel pain don't they? We should consider their lives too. What about plants?

C: Effect on plants. Well I am no botanist, but do plants feel pain? If not, why consider them?

P: Plants have interests. And we depend on them. What about rivers, streams, the land itself?

C: Well, I am not really sure. I guess the toxins produced in the explosion had to go somewhere. Maybe they are bioaccumulating up the food chain. I guess we will need to ask experts and do more research.

P: What about air? What is air, anyway?

C: Air is made up of molecules of N_2, O_2, Ar, CO_2 and gaseous H_2O. In general clean air doesn't contain toxic or dangerous chemicals.

P: What do all of those symbols mean? It all sounds Greek to me! Where would I go to learn about dangerous chemicals?

C: Well there is the TRI, MSDS and NIMBY information.

P: More acronyms I do not know! Why are we creating pesticides that are known to be dangerous?

C: We need to kill pests to maximize crop yields. If a plague of locusts eat your entire crop, what then?

P: How can a molecule be benign and deadly at the same time?

C: Well, there can be unanticipated effects due to the chemical behaving in ways that are not in the original design. Do you remember Thalidomide? The molecule is chiral and one enantiomer is a great remedy for nausea while the other causes birth defects.

P: Enantiomer? What's the relation between Bhopal and clothes?

C: Well, what are pesticides used for?

P: Cotton. But why was the plant in India in the first place?

C: Well the chemicals can be made anywhere. Isn't the labor cheaper in India? Why not make the dangerous chemicals near where they are used rather than transporting them?

C: Is there a way to calculate the costs and benefits of manufacturing a pesticide like MIC?

P: Economists use cost/benefit analyses; philosophers push for full cost accounting.

C: What do we do when the costs outweigh the benefits? Who decides in the end? Why did the Indian government prevent the US company from inspecting and maintaining the plant? Why did they have to run the plant completely? What relationship problems between the US and India fueled this?

The questions in the preceding dialogue can be categorized into at least three different areas: topical (integrative seminar), philosophical (environmental ethics) and chemical (fundamental chemistry in context). These questions would serve to drive the content for all three areas of the learning community in the first week of the module. Upon investigating the answers to these questions, however, students are still left to consider what to do with the knowledge they have gained. The content of the second week attempts to introduce the different approaches that could be taken to mitigate the problems identified in the first week. One could envision any number of approaches including those centered on green chemical, ethical, governmental, regulational, economical, or from a global or public health viewpoint. While the ideal learning community would involve all theoretical or applied approaches, our learning community focused upon the former two. The content to be discussed in the first two-week learning community module is visually outlined in Table Va; questions to drive the content in the first two-week learning community module are visually outlined in Table Vb.

The two-week structure of the topical integrative seminar dictates the chemical and ethical content required in each of the courses. The generic two-week structure including both on-line and face-to-face meetings of the integrative seminar can be visualized in Table VI.

Although on the surface it may appear daunting to coordinate content simultaneously for all three segments of the learning community, it is possible with the help of an online learning environment. The topical questions, reading assignments, and course assignments for both the online and face-to-face environments for each week in the first two-week modules are provided in Figures 1 and 2. The initial design for each of the modules was to focus on "a problem" in the first week while the second week focuses upon "solutions." This can be clearly observed in the first and second "overview-page" shown for General Consumption Module in Figures 1 and 2.

At the conclusion of each module, students collectively write a formal paper that captures the community's progression of thought and synthesizes the

Table Va. Content Addressed by Questions in the First Two-Week Learning Community Module for Conscience and Consumption

<table>
<tr><td colspan="4">General Consumption module

Week 1: What is the impact of our general consumption on the environment?
Bhopal Case Study & Closet Investigation

Week 2: How does green consumption lessen the impact on the environment?</td></tr>
<tr><td colspan="2">Environmental Ethics

Animal rights
The Land Ethic
Intrinsic, instrumental and systemic value
Cost-benefit analysis
Full cost accounting</td><td colspan="2">Chemistry

What is - a chemical? a mixture? a compound? a molecule? an element?; Safety of chemicals.
Scientific notation, conversions; Periodic Table of Elements; Naming simple compounds; Atomic structure and periodicity; Balancing equations; Drawing molecules; Interaction of light with molecules; Chlorofluorocarbons (CFC's) and ozone; Development of green pesticides</td></tr>
<tr><td>Global Issues</td><td>Economic Incentives</td><td>Governmental Regulation</td><td>Public Health</td></tr>
</table>

community's conclusions concerning a) the most effective approaches to minimizing the impact of our specific consumption on the environment, b) specific projects that could to bring about change. Students communally author the paper through an online Wiki.

All scheduled class time in the final two weeks of the course is dedicated completely to the civic engagement projects. Over the course of the first week, students develop a project idea within a group, develop a list of questions they need to answer to explain and justify the plan; develop a plan of action to fill in the structure for advocacy and begin to carry out the plan. Project ideas can be but are not required to be derived from the specific projects that were previously outlined in the Wiki earlier in the semester for each module. At the beginning of the second week, each group defends their plan idea and research to the rest of the class in a debate forum. Final project outcomes will vary, but each group is responsible for composing a project summary article to the student newspaper, a 15-minute power-point presentation for a general audience as well as a poster

Table Vb. Questions to Drive the Content in the First Two-Week Learning Community Module for Conscience and Consumption

<table>
<tr><td colspan="2">General Consumption module

Week 1
What is the impact of our general consumption on the environment? What is in your closet and why? What is the connection between Bhopal and clothes? What happened at Bhopal?

Week 2
How does green consumption lessen the impact on the environment? How can we begin to minimize the impact of our lifestyle on the environment</td></tr>
<tr><td>Environmental Ethics

Week 1
Who, or what, is worthy of moral consideration? Why do we create pesticides that are known to be dangerous? What was disastrous about the disaster at Bhopal? Why did the chemical reaction occur at Bhopal? What value is there to cost-benefit analyses of proposed preventative measures or solutions?

Week 2
What are the major pitfalls in applying cost-benefit analyses to proposed preventative measures or solutions? Should we begin to minimize our impact of our lifestyle on the environment in the future?</td><td>Chemistry

Week 1
What are the fundamental building blocks of our shoes, ourselves, our consumer goods our universe? What are major air pollutants we encounter every day and what effects can they have on us? What is air? What is clean air? What is polluted air? Why is methylisocyanate so harmful? How can a molecule be benign and deadly at the same time?

Week 2
Are there ways of preserving lifestyle and patterns of consumption without negatively impacting the environment?</td></tr>
</table>

Table VI. Integrative Seminar: Generic Structure for Each Two-Week Segment[a]

Pre Seminar (Online)[b]	*Seminar (Face-to-Face)*	*Post Seminar (Online)*[b]
A: Pre-IS#1 Assigned Readings; Online Research; Questions to Guide the Readings and Research centered around a question that requires the knowledge of chemistry and consideration of ethical issues.	***B: IS Week #1*** Discuss our research results. Continuation of the online discussion. Answer questions, pose new questions. What do/don't we know? How do we find out? Where do we go from here?	***C: Post-IS#1 / Pre-IS#2*** Follow-up posts; reflections and introduction to new questions raised by seminar discussion.
C: Post-IS#1 / Pre-IS#2 New Readings, Questions, Postings, Working toward a chemically benign/ green, ethically justifiable solution.	***D: IS Week #2*** Present and discuss research results	***E: Post-IS#2*** Write a formal paper that captures the community's progression of thought and synthesizes the community's conclusions concerning a) the most effective approaches to minimizing the impact of our consumption on the environment, b) specific projects that could to bring about change.

a. Note that C: Post-IS#1 / Pre-IS#2 includes both follow-up after the Integrative Seminar #1 face-to-face meeting and online discussion in preparation for Integrative Seminar #2.

b. Online environment was developed with help from a Pottruck Technology Resource Center Faculty Fellowship Grant.

What is the impact of our general consumption on the environment?			
Week two	Ethics	Chemistry	Integrative Seminar
Sun			**What's in your closet?** Online Reading: Cleaning the Closet by Schorr Do: Closet Exercise "What's in your closet?" Online Reading: "Legacy of Bhopal" Case Study Post 2.1: By Monday Morning 9 am.
Mon			**What is the connection between Bhopal and clothes?** Post 2.2: What is the connection between Bhopal and clothes? Hint" What is a large export cash crop in India in the 1980's?
Tues	**Who, or what is worthy of moral consideration?** Read: All Animals are Equal, Singer Read: The Land Ethic, Aldo Leopold Online Reading: An Animal's Place, Poltan Journal 2.1: What is speciesism, and why do some philosophers think it is objectionable?	**What is Air?** Read: Sections 1.1 – 1.8 in Chemistry in Context. Do: Begin Week 2 Homework	**What is the connection between Bhopal and clothes?** Respond 2.2: to two of your classmates postings.
Wed		**What is clean air? What is polluted air? (Laboratory)** Read: the experimental handout. Postlab: Answer post-lab questions and submit data sheet before Next Monday at 9 am.	**Why was Union Carbide in Bhopal in the first place? Why were they making pesticides that were known to be dangerous?** Reply 2.2: to your classmates' questions or comments in their Response:
Thurs	**What was disastrous about the disaster at Bhopal?** **Online Reading:** Value in Nature and the Nature of Value, Ralston Read: Animal Liberation and Environmental Ethics: Bad Marriage, Quick Divorce, Mark Sagoff Journal 2.2: Sagoff argues that "a humanitarian ethic – an appreciation not of nature, but of the welfare of animals will not help us to understand or to justify an environmental ethic" (p 43). What reasons does he give for this view? Do you agree?	**What are the fundamental building blocks of our shoes, selves, consumer goods and universe?** Read: Sections 1.9 – 1.14 in Chemistry in Context Do: Complete Week 2 Homework before Quiz. Quiz Week 2: Complete before Monday at 9am.	**How harmful is methylisocyanate?** Do: Locate and review the Materials Safety Data Sheet for methylisocyanate. Write: Combine your research results into a summary with references that will serve as the foundation for your presentation to the community on Friday.
Fri			**What is the impact of our lifestyle on the environment now?** Discuss: Discuss our research results. Continuation of the online discussion. Answer questions, pose new questions. What do/don't we know? How do we find out? Where do we go from here?

Figure 1. Overview of the first week schedule for general consumption module (See reference 46.)

How does green consumption lessen the impact on the environment?			
Week three	Ethics	Chemistry	Integrative Seminar
Sun			**Can we begin to minimize the impact of our lifestyle on the environment in the future?** Online Reading: We all Live in Bhopal, George Bradford Online Reading: Development of green pesticides: Confirm, Mack 2 and INTREPID Online Reading: Sentricon: Termite Control, Joshua Fung
Mon			**What different approaches might we take to minimize the impact of our lifestyle on the environment in the future?** Post 3.1: Think broadly about all of the different approaches we might take to minimize the impact of our lifestyle on the environment. A scientific and technological approach is obviously one; what others can you think of? Provide examples if possible to support your claims. The following recent articles about Bhopal from five different authors may spark your thoughts: Crabb, Hertzgaard, Reily, Sanghi, Sharma.
Tues	**What value is there to CBA of proposed preventative measures?** Online Reading: Chapters 6-8, Environmental Economics, Field & Field. Journal 3.1: What costs and benefits should we include in a CBA of the clean up of Bhopal.	**How can a molecule be benign and deadly at the same time?** Read: Sections 2.1-2.7 in Chemistry in Context. Quiz 3.1: Complete before Thursday.	**What different approaches might we take to minimize the impact of our lifestyle on the environment ?** Respond 3.1: to two of your classmates postings. You might need to do additional research to support your suggestions. Be sure to follow the guidelines for a good discussion post,
Wed		**How do we know pollutants are in the air? How do we measure the pollutants in the air?** Read: the experimental handout. Do: Answer the post-lab questions	**How effective do you think each of these different approaches could be?** Reply 3.1: to your classmate's questions or comments in their RESPONSE.
Thurs	**What are the major pitfalls in applying cost-benefit analyses to proposed preventative measures?** Online Reading: Chapter 9, Environmental Economics, Field & Field. Read: A Place for Cost-Benefit Analysis, Schmidtz. Journal 3.2: Is CBA a reasonable way to address environmental concerns? Explain and defend your answer. Write: Write a 4-5 page reflection paper: What matters in environmental ethics? Due one week from next Tuesday.	**How is methyl isocyanate analogous to ozone?** Read: Sections 2.7 – 2.12 in Chemistry in Context. Quiz 3.2: Complete before Saturday at 2pm.	**What different approaches will allow us to be most effective in preserving lifestyle and patterns of consumption without negatively impacting the environment globally?** Write: Combine your research results and thoughts over the past two weeks into a summary with references that will serve as the foundation for your presentation to the community on Friday.
Fri			**What specific projects could be launched at Simmons College to bring about change?** Do: Present your research results. By Monday at Midnight: Do Group on Wiki: Write a formal paper that captures the community's progression of thought and synthesizes the community's conclusions concerning a) the most effective approaches to minimizing the impact of our consumption on the environment, b) specific projects that could be launched at Simmons College to bring about change.

Figure 2. Overview of the second week schedule for the general consumption module.

presentation to be presented at the Undergraduate Research Conference at the conclusion of the academic year.

A summary of the case studies, personal exercises and course content for the environmental ethics and chemistry courses as paced to each of the two-week topic modules is outlined in Table VII. The course as outlined above will be offered in the Fall Semester 2009 at Simmons College.

Conclusions

Our experiences over the past seven years indicate that Green Chemistry lecture based courses encourage and inspire individuals to civic action. Herein, three distinctly different green chemistry lecture-based courses were detailed and the design and outcome of the corresponding civic-engagement projects were described. Students enrolled in a science-major, senior-seminar course organized a day-long "Undergraduate Green Chemistry Symposium" for the scientific community at Northwestern University and a week long "Green Chemistry Exhibition" at Simmons College. Students enrolled in a non-science major topical Honors Program course researched green consumer products and informed the community by submitting a weekly Green Chemistry article to the student newspaper and by creating a green version of the "What to bring to college" guide for incoming students. From these experiences we have designed and proposed what we believe will be the most effective method to educate non-science majors in green chemistry: a learning-community course coupling green chemistry and environmental ethics.

Acknowledgements

The authors acknowledge financial support from the Women In Materials Program at Simmons College funded through the National Science Foundation, Directorate for Mathematical & Physical Sciences, Divisions of Materials Research and Chemistry under grant number DMR-0108497, Simmons College for a New Faculty Start-up Grant, the Henry and Camille Dreyfus Foundation Postdoctoral Fellowship in Environmental Chemistry, the Merck/AAAS Undergraduate Science Research Program and a Pottruck Technology Resource Center Faculty Fellowship Grant. Contributions from students enrolled in CHM 393 at Northwestern University (Winter 2002, 2003) and HON 302 at Simmons College (Fall 2004, Fall 2005), the Departments of Chemistry and Philosophy as well as the Honors Program and the PTRC at Simmons College are gratefully acknowledged. Special thanks is extended to Emily Scott-Pottruck, Mary Jane Treacy, Don Basch, Connie Chow, Tania Cabrera, Matthew Skaruppa, Kathryn Gerth, Natalie Kaufman, Son-Binh Nguyen, Joseph Hupp, Braddlee, Gail Mathews-DeNatale, Sean Wright and Jason Gorman.

Table VII. Summary of the Case Studies, Personal Exercises, and Course Content for the Environmental Ethics and Chemistry Courses

Module *Case Study* Personal Exercise	*Environmental Ethics Content*	*Chemistry Content*
General Consumption *Bhopal* Closet Investigation	Animal rights; The Land Ethic; Intrinsic, instrumental and systemic value; Cost-benefit analysis; Full cost accounting	What is - a chemical? a mixture? a compound? a molecule? an element? Chemical safety; Scientific notation, conversions; Periodic Table of Elements; Naming simple compounds; Atomic structure and periodicity; Balancing equations; Drawing molecules; Interaction of light with molecules; Chloro-Fluoro-Carbons (CFC's) and ozone; Development of green pesticides
Oil Consumption *Exxon Valdez* Materials With Which You Come in Contact	Moral standing vs legal standing; Needs, interests, and rights; Latent tendencies; Right for plants? Rights for species? Respect for nature; Biocentrism vs anthropocentrism	Carbon dioxide emissions; Green house effect; Carbon cycle; Shapes of molecules; Energy; Life-cycle analysis calculations; Energy content in fuels
Water Consumption *Chlorine Sunrise* Water Use Activity	Individualism; Holism; Deep ecology; Social ecology; Intrinsic, instrumental and systemic value again; Aesthetic value	Water as a solvent; Concentrations: molarity; Solutions; Hydrogen bonding; Ionic vs covalent; Water contaminants; Acids, bases, pH; Consumer sources of chlorine; Chlorine alternatives
Energy Consumption *Chernobyl* Trash Bag Activity -How do we clean up a mess?	The value of wilderness; Preservation; Restoration; Conservation; Nature/culture dualism; Domination of nature or benevolent restoration? Moral character; Virtue ethics. Land virtues	Clean nuclear energy? Nuclear fission / fusion; Radioactivity; Isotopes; Nuclear waste; Alternative energy sources; Electron transfer; Batteries / fuel cells; Hydrogen economy; Photovolatics
Plastic Consumption *The PVC Disaster* Paper vs Plastic Exercise	Moral Vice; Gluttony, greed, apathy, arrogance; What is the good life? Do we consume too much? Environmental pragmatism; Activism; Role of philosophy in promoting change.	Plastics and polymers -revolution or nightmare? Catalysts; The "Big Six" polymers; Introduction to organic chemistry; Physical properties; Closed vs open loop recycling; Biodegradable? Cradle-to-Cradle; Plastics - shades of green; NatureWorks™ PLA.

References

1. Anastas, P. T.; Warner, J. C. *Green Chemistry: Theory and Practice*; Oxford University Press: Oxford, U.K., 1998.
2. The Greener Education Materials (GEMs) for Chemists Database. http://greenchem.uoregon.edu/gems.html (accessed Apr 19, 2008).
3. Cann, M. C.; Connelly, M. E. *Real-World Cases in Green Chemistry*; American Chemical Society: Washington, DC, 2000.
4. *Green Chemistry: Frontiers in Benign Chemical Synthesis and Process*; Anastas, P. T., Williamson, T. C., Eds.; Oxford University Press: Oxford, U.K., 1998; Chapter 1.
5. (a) Breslow, R. *Chem. Eng. News* **1996**, *74(35)*, 72. (b) Morrissey, S. *Chem. Eng. News* **2002**, *80(20)*, 46. (c) Wilkinson, S. L. *Chem. Eng. News* **1997**, *75(41)*, 35-43. (d) Lancaster, M. *Educ. Chem.* **2000**, March, 40-43; (e) Clark, J. H. *Green Chem.* **1999**, *1*, 1-8.
6. Cann, M. C.; Connelly, M. E. *Real-World Cases in Green Chemistry*; American Chemical Society: Washington, DC, 2000; Chapter 1. (b) Trost, B. M. *Science* **1991**, *254*, 1471-1477. (c) Trost, B. M. *Acc. Chem. Res.* **2002**, *35*, 695-705.
7. Cann, M. C.; Connelly, M. E. *Real-World Cases in Green Chemistry*; American Chemical Society: Washington, DC, 2000; Chapter 3.
8. (a) Borman, S. *Chem. Eng. News* **2002**, *80(6)*, 29-34. (b) Kolb, H. C.; Finn, M. G.; Sharpless, K. B. *Angew. Chem., Int. Ed.* **2001**, *40*, 2004-2021.
9. Norman, W.; MacDonald, C. "Triple Bottom Line"—a Critique. http://www. businessethics.ca/3bl/ (accessed Apr 19, 2008).
10. Cann, M. C.; Connelly, M. E. *Real-World Cases in Green Chemistry*; American Chemical Society: Washington, DC, 2000; Chapter 4.
11. Hooper, R.; Potter, A. K. N.; Singh, M. M. *Green Chem.* **2001**, *3*, 57-60.
12. Plastics 101. http://www.americanchemistry.com/s_plastics/ sec_learning. asp?CID=1571&DID=5957 (accessed Apr 19, 2008).
13. (a) Allen, B. *Green Chem.* **1999**, *1*, G142-G143. (b) Lam, K. K.; Ottewill, G.; Plunkett, B.; Walsh, F. *Green Chem.* **1999**, *1*, G10-G109. (c) University of California Toxic Substances Research & Teaching Program. Health and Environmental Assessment of MTBE. http://www. tsrtp.ucdavis.edu/public/mtbe/mtberpt/index.php (accessed Apr 19, 2008).
14. (a) Rocha, F. R. P.; Nóbrega, J. A.; Filho, O. F. *Green Chem.* **2001**, *3*, 216-200. (b) Csihony, S.; Mehdi, H.; Horváth, I. T. *Green Chem.* **2001**, *3*, 307-309.
15. U.S. Environmental Protection Agency. Toxics Release Inventory (TRI) Program Web Site. http://www.epa.gov/tri (accessed Apr 19, 2008).
16. (a) Curzons, A. D.; Constable, D. J. C.; Mortimer, D. N.; Cunningham V. L. *Green Chem.* **2001**, *3*, 1-6. (b) Constable, D. J. C.; Curzons, A. D.; Freitas dos Santos, L. M.; Geen, G. R.; Hannah, R. E.; Hayler, J. D.; Kitteringham,

J.; McGuire, M. A.; Richardson, J. E.; Smith, P.; Webb, R. L.; Yu, M. *Green Chem.* **2001**, *3*, 7-9.

17. (a) van Vliet, M. C. A.; Mandelli, D.; Arends, I. W. C. E.; Schuchardt, U.; Sheldon, R. A. *Green Chem.* **2001**, *3*, 243-246. (b) Smith, K.; He, P.; Taylor, A. *Green Chem.* **1999**, *1*, 35-38. (c) Clark, J. H.; Macquarrie, D. J. *Chem. Commun.* **1998**, 853-860.
18. Braddock, C. *Green Chem.* **2001**, *3*, G26-G32.
19. (a) Choudary, B. M.; Kantam, M. L.; Neeraja, V.; Rao, K. K.; Figueras, F.; Delmotte, L. *Green Chem.* **2001**, *3*, 257-260. (b) Brown, A. S. C.; Hargreaves, J. S. J. *Green Chem.* **1999**, *1*, 17-20. (c) Yadav, G. D.; Pujari, A. A. *Green Chem.* **1999**, *1*, 69-74.
20. Anastas, P. T.; Kirchhoff, M. M.; Williamson, T. C. *Appl. Catal., A* **2001**, *221*, 3-13.
21. Cann, M. C.; Connelly, M. E. *Real-World Cases in Green Chemistry*; American Chemical Society: Washington, DC, 2000; Chapter 9.
22. Curran, D.; Lee, Z. *Green Chem.* **2001**, *3*, G3-G7.
23. (a) Holbrey, J. D.; Seddon, K. R. *Clean Products and Processes* **1999B**, *1*, 223-236. (b) Earle, M. J.; Seddon, K. R. *Pure Appl. Chem.* **2000**, *72*, 1391-1398.
24. (a) Cann, M. C.; Connelly, M. E. *Real-World Cases in Green Chemistry*; American Chemical Society: Washington, DC, 2000; Chapter 2. (b) Kirchhoff, M. *ChemMatters* **2000**, *18* (2), 14-15.
25. Nelson, W. M. *Green Solvents for Chemistry: Perspectives and Practice*; American Chemical Society: Washington, DC, 2003.
26. Frost, J. W.; Lievense, J. *New J. Chem.* **1994**, *18*, 341-348.
27. (a) Jones, D. *ChemMatters* **2000**, *18* (4), 4-6. (b) Nice, K.; Strickland, J. How Fuel Cells Work. http://auto.howstuffworks.com/fuel-cell.html (accessed Apr 19, 2008).
28. (a) Bozell, J. J.; Hobers, J. O.; Claffey, D.; Hames, B. R.; Dimmel, D. R. In *Green Chemistry: Frontiers in Benign Chemical Synthesis and Process*; Anastas, P. T., Williamson, T. C., Eds.; Oxford University Press: Oxford, U.K., 1998; Chapter 2, pp 27-43. (b) Bartle, I.; Oliver, N. *Green Chem.* **1999**, *1*, G6-G9.
29. (a) Conversion of Biomass Wastes to Levulinic Acid. http://www.pnl.gov/biobased/docs/mthf.pdf (accessed Apr 19, 2008). (b) U.S. Environmental Protection Agency. 1999 Small Business Award. http://www.epa.gov/greenchemistry/pubs/pgcc/winners/sba99.html (accessed Apr 19, 2008).
30. (a) Fukuda, H.; Kondo, A.; Noda, H. *J. Biosci. Bioeng.* **2001**, *92*, 405-416. (b) Haas, M. J.; Scott, K. M.; Alleman, T. L.; McCormick, R. L. *Energy Fuels* **2001**, *15*, 1207-1212.
31. (a) Horeis, G.; Pichler, S.; Stadler, A.; Gössler, W.; Kappe, C.O. Microwave-Assisted Organic Synthesis—Back to the Roots. Presented at the Fifth International Electronic Conference on Synthetic Organic

Chemistry [Online], September 1-30, 2001; Paper e0000. (b) Varma, R. S. *Green Chem.* **1999**, *1*, 43-55.

32. Hudlicky, T.; Frey, D. A.; Koroniak, L.; Claeboe, C. D.; Brammer, L. E., Jr. *Green Chem.* **1999**, *1*, 57-59.
33. Schiel, C.; Oelgemöller, M.; Ortner, J.; Mattay, J. *Green Chem.* **2001**, *3*, 224-228.
34. (a) Bashkin, J.; Rains, R.; Stern, M. *Green Chem.* **1999**, *1*, G41-G43; Cann, M. C.; Connelly, M. E. *Real-World Cases in Green Chemistry*; American Chemical Society: Washington, DC, 2000; Chapters 6-8 and 10.
35. Eubanks, L. P.; Middlecamp, C. H; Pienta, N. J.; Heltzel, C. E.; Weaver, G. C. *Chemistry in Context*, 5th ed.; McGraw-Hill: New York, 2006.
36. McDonough, W.; Braungart, M. *Cradle to Cradle: Rethinking the Way We Make Things*; North Point Press: New York, 2002.
37. Newton, L. H.; Dillingham, C. K. *Watersheds: Classic Cases in Environmental Ethics*; Thomson Wadsworth: Belmont, CA, 1994; Chapter 4.
38. Newton, L. H.; Dillingham, C. K. *Watersheds: Classic Cases in Environmental Ethics*; Thomson Wadsworth: Belmont, CA, 1994; Chapter 7.
39. Newton, L. H.; Dillingham, C. K.; Choly, J. *Watersheds 4: Ten Cases in Environmental Ethics*; Thomson Wadsworth: Belmont, CA, 2006; Chapter 1.
40. Newton, L. H.; Dillingham, C. K.; Choly, J. *Watersheds 4: Ten Cases in Environmental Ethics*; Thomson Wadsworth: Belmont, CA, 2006; Chapter 9.
41. Newton, L. H.; Dillingham, C. K.; Choly, J. *Watersheds 4: Ten Cases in Environmental Ethics*; Thomson Wadsworth: Belmont, CA, 2006; Chapter 7.
42. Newton, L. H.; Dillingham, C. K. *Watersheds: Classic Cases in Environmental Ethics*; Thomson Wadsworth: Belmont, CA, 1994; Chapter 3.
43. Newton, L. H.; Dillingham, C. K.; Choly, J. *Watersheds 4: Ten Cases in Environmental Ethics*; Thomson Wadsworth: Belmont, CA, 2006; Chapter 4.
44. Newton, L. H.; Dillingham, C. K.; Choly, J. *Watersheds 4: Ten Cases in Environmental Ethics*; Thomson Wadsworth: Belmont, CA, 2006; Chapter 2.
45. Allegheny County Health Department. A Consumers Guide To Reducing Mercury Pollution *And* Exposure. http://www.achd.net/airqual/pubs/pdf/mercury04.pdf (accessed Apr 19, 2008).
46. All readings are from Schmidtz, D; Schmidtz, E. *Environmental Ethics: What Really Matters, What Really Works*, Oxford University Press: Oxford, U.K., 2001.

Chapter 5

Integrating Green Chemistry into the Introductory Chemistry Curriculum

Marc A. Klingshirn[1] and Gary O. Spessard[2]

[1]Department of Chemistry, University of Illinois at Springfield, Springfield, IL 62703
[2]Chemistry Department, St. Olaf College, Northfield, MN 55057

Green chemistry education offers a solution to our current environmental problems because it provides the opportunity to train future scientists and political leaders, thus helping move us toward a more sustainable society. Green chemistry, while becoming more commonplace in today's curricula, has seen the greatest degree of implementation in the organic chemistry laboratory. It is only recently that introduction of green chemistry principles into the first year chemistry courses has been addressed. This in spite of the need for such education to be uniform throughout a student's chemistry curriculum from the beginning courses onward. Successful case studies and examples of implementation of green chemistry into the lecture and laboratory of first-year courses will be covered. Two redesigned experiments relating to the formula of a hydrate and metal complexation will be discussed in addition to key drivers and major barriers to green chemistry implementation.

Introduction

Today's society is becoming more and more environmentally conscious. With the threat of global climate change and the call for movement toward more sustainable practices, the world is being faced with tough challenges (*1-3*). We, as a society, must change our ways, educate our children, strive to be more aware of our personal actions, and think on a larger scale than that involved with our immediate surroundings. Green chemistry can play an integral role in moving society toward a more positive, sustainable direction.

Implementation of green chemistry education at the undergraduate level is key to adopting more sustainable practices and hence a more sustainable society (*4*). This education should be for both science and non-science majors. While certainly those students who are science majors may have a direct route to action in applying green chemistry principles in both research and development, non-science majors can also benefit since they become aware of the importance of sustainable practices, learn about positive every-day habits, and come to realize that the chemical industry, which in the past (and even currently) was viewed as a major contributor to environmental degradation, is also part of the solution to the problem.

A second driver that pushes the development of green chemistry is economics. While the goal of most industries is to provide the public with goods and services that help better lives, each industry regardless of the type must also maximize profit margins. Many companies are realizing that green chemistry offers financial advantages (*5-8*). Since the practice of green chemistry involves process and energy efficiency as well as reduction of waste, one outcome is financial savings through development of shorter synthesis processes, reduction of energy costs, and lowering the need for waste disposal.

To further support green chemistry research, companies such as Pfizer give awards to their own scientists who develop "a new practice, technology, or project that advances the principles of green chemistry in Pfizer, and which also maintains or enhances productivity." (*9*) A "quantifiable environmental benefit" must also be demonstrated. The award also recognizes university chemistry departments that pursue green chemistry. Support of these universities by large corporations such as Pfizer indicates that these firms are seriously looking at the educational pipeline as a future source of scientists who are capable of applying green chemistry principles to chemical processes.

The bottom line in this discussion is that our collective society, including industry, must make conscious decisions regarding our environment. The easiest way to ensure that these decisions focus on sustainability is through support of green chemistry implementation into education at all levels, beginning in the middle school, progressing through high school, and further continuating throughout the entire college curriculum.

Most work in green chemistry education implementation has focused on the undergraduate curriculum (*10*). While the progressive movement to include and discuss the importance, practice, and principles of green chemistry in the undergraduate setting is encouraging, implementation across the curriculum is by no means uniform. The greatest progress has been seen in the organic chemistry laboratory with little to slow progress at the first-year level (*11,12*). This is beginning to change. With concerted efforts by the American Chemical Society's Green Chemistry Institute (*13*), dedicated faculty, and the presence of forward-thinking chemistry programs in some colleges and universities in the United States and throughout the world, momentum is beginning to build. While the argument could be made that the principles of green chemistry are most applicable in organic chemistry, because of its focus on synthesis, the first-year chemistry course is a prime venue to discuss concepts such as atom economy, energy efficiency, chemical hazards, and general laboratory safety (*14*). By exposing the students to these topics early in their careers, the tenets of green chemistry can become more concrete as they are encountered in subsequent courses. In this chapter we will discuss the current status of green chemistry implementation into the first-year curriculum, and we will describe successful tactics for incorporating green chemistry into both the traditional and laboratory classroom. We will also address other aspects such as student feedback, possible future directions, and obstacles faced during implementation.

A question that educators often pose is how to effectively incorporate green chemistry education into an already crowded curriculum. While it certainly is true that we are always re-evaluating our course content to cover the essentials and keep it relevant, this does not mean that we should feel limited in our ability to incorporate green chemistry. In many cases green chemistry offers a better alternative to traditional material in getting across the principles covered in the first-year curriculum. Thus green chemistry should not be perceived as an "add-on" topic but instead should replace some material that is traditionally covered.

In the Classroom

Green chemistry is most easily demonstrated in the laboratory since the theory is put into action; however, integration should not be isolated only to the laboratory. As with any chemical theory, the best way to achieve student comprehension is to discuss the theory and then allow implementation in some form. In addition to laboratory exercises, other means of exposure to green chemistry principles can take the form of papers, group projects, or activities that allow students to relate chemistry to societal problems. Quite often the most effective method is to have the students make personal connections between a chemical topic and their own lives. It is only then that they begin to understand the relevance and importance of green chemistry in everyday life.

Green Chemistry Exercises

In the mid-1990s, Carnegie Mellon University's Environmental Institute began to implement the "Environment across the Curriculum" initiative (*15*). The goal was to develop modules that contained environmental themes that could be used "in courses in every college in the university..." One of the modules, which focused on green chemistry and was specifically designed for the first year chemistry student, will be discussed.

This module consisted of a three-lecture sequence that outlines the history of refrigeration with discussions on original chemicals/technology used and the progressive change to more environmentally friendly alternatives. The "Greening of Refrigeration" module effectively demonstrates how proven chemical technologies can ultimately be harmful to the environment, but also how environmentally sound chemical research can be a solution to the problem.

More recently, Mike Cann of the Chemistry Department at the University of Scranton developed green chemistry modules that can be implemented into traditional chemistry courses such as general, organic, inorganic and biochemistry as well as non-traditional offerings including chemical toxicology and industrial chemistry (*16*). The module "Design and Application of Surfactants for Carbon Dioxide; Making Carbon Dioxide a Better Solvent in an Effort to Replace Solvents that Damage the Environment" is specifically designed for a general chemistry audience. It is suitable for discussions centering on phase diagrams, supercritical fluids, and molecular polarity. These modules can be found at the University of Scranton's Chemistry Department's website.

Song and co-workers developed a set of exercises that allows students the opportunity to analyze a series of reactions and judge them based on choice of feedstock, atom economy, reaction conditions, environmental exposure, and resource conservation (*17*). One example asks the students to consider the following two series of reactions and then decide how to synthesize eight moles of aluminum hydroxide (Figure 1).

$$Al \xrightarrow{H_2SO_4} Al_2(SO_4)_3 \xrightarrow{NaOH} Al(OH)_3$$

$$Al \xrightarrow{NaOH} Na[Al(OH)_4] \xrightarrow{H_2SO_4} Al(OH)_3$$

Figure 1. Alternate syntheses of aluminum hydroxide.

The students were then asked to answer the following questions:

1. Which reaction is better in terms of using less feedstock?
2. Propose a new reaction that can save more chemicals.

The students ultimately find that the second reaction scheme utilizes fewer moles of sodium hydroxide and sulfuric acid. Exercises such as this not only help teach about green chemistry principles, but also they teach about fundamental chemical concepts such as balancing chemical equations, stoichiometry, and the mole concept.

Presidential Green Chemistry Awards

President Clinton established the Presidential Green Chemistry Award in 1995. The goal of the award was to honor those in business, academics, and other organizations for their contributions in the development of environmentally-friendly chemical technologies. These technologies were developed with the twelve principles of green chemistry as the cornerstone of the work. These awards and the chemistry behind them provide rich examples for integration of green chemistry into both upper-level (*18*) *and* first-year courses. Not only do the awarded projects teach about cutting-edge technology and research, they also put chemistry into context since many of them involve technology that is familiar and relevant to students.

The short film "Green Chemistry: Innovations for a Cleaner World," produced by the American Chemical Society, also highlights recent EPA Presidential Green Chemistry Award winners. In a past course taught at St. Olaf College, this film was shown in class to give the students additional information on the type of green chemistry research that is currently taking place in both industry and academia. The students were then asked to write a short research paper on a topic of their choice with the only requirement being that the subject matter needed to have an environmental theme. Many students used this movie as a "spring board" for ideas for their essays. A few student essay topics included:

- Sustainability issues in the hospital environment,
- The Biogas Support Program and its use in Third World countries,
- The anti-cancer drug Taxol and how green chemistry has influenced its synthesis,
- Biodiesel and ethanol as viable fuel alternatives, and
- Green chemistry and its application of the production of 'greener' plastics.

To obtain student opinion on the research paper, an additional question relating to it was added to end-of-the-term course evaluation. Student response to the research paper was very favorable, since 79.2% felt the essay on green chemistry/environmental issues was a worthwhile and useful learning experience.

In the Laboratory

The laboratory is the easiest place for students to begin understanding what green chemistry is and how it can be implemented. While the traditional lecture or discussion setting is useful for addressing the principles and theory of green chemistry, students often do not begin to critically think about the theory until it is put into practice. The lectures are often viewed as abstract by students while the laboratory setting is perceived as more applied and concrete.

The easiest methods of incorporation of green chemistry into the laboratory curriculum involve little-to-no effort other than the occasional modification to an existing lab or switching a lab for one that is "greener". With resources such as the *Journal of Chemical Education* and the University of Oregon's GEMS database (Greener Education Materials for Chemists) (*19*) the ability to locate teaching materials is becoming easier. Also now available is The Green Chemistry Education Network (GCEdNet) (*20*), a new clearinghouse of information on green chemistry with a focus on disseminating new greener laboratory exercises.

Methods of Incorporation

Relatively simple tasks can be asked of students in a laboratory period to help them grasp the concepts of green chemistry. Since green chemistry advocates the use of environmentally benign, non-toxic reagents, and the use of reagents that stem from renewable resources, asking students to consult Material Safety Data Sheets (MSDS) for the chemicals that they are using is an effective means of teaching students about the hazards associated with a specific chemical. Another alternative is for the students to access them using online databases. While the amount of information contained within an MSDS can be overwhelming, having the students focus on specific aspects such as the National Fire Protection Association (NFPA) ratings, permissible exposure limits (PEL), and handling precautions makes the process less intimidating.

Integration of green chemistry principles into pre-lab and post-lab questions is another easy and effective inclusion method. The synthesis of aspirin is a common first-year lab procedure. One of the fundamental principles of green chemistry is atom economy. A possible pre-lab question might be to have the students calculate the percent atom economy of the synthesis using the following equation:

$$\%\,AtomEconomy = \frac{atomic\ wt.\ of\ atoms\ in\ the\ useful\ product}{sum\ of\ atomic\ wts.\ of\ all\ starting\ materials} x\,100\%$$

This relatively simple calculation can then lead to questions about whether the atom economy calculated from the balanced equation is truly achievable, and if not, how that relates to the issues of waste.

Post-lab questions can be generated that force students to compare and contrast old procedures with newer, greener syntheses and determine the relative advantages and disadvantages of each. For example if a laboratory procedure has been redesigned to substitute the strong oxidizer chromium(VI) with 3% hydrogen peroxide, post-lab questions could be included that ask students to consult the MSDSs of each chemical and list the NFPA ratings, chemical properties, PEL, and so on.

Waste analysis/monitoring can also be used to help students grasp the concept of the amount of waste generated. Green chemists strive to develop chemical processes that generate little to no waste or allow for recycling of materials back into the system to create a closed-loop system. If recycling cannot be achieved, requiring the students to keep a running tabulation of the amount of each type of chemical used allows the students to begin to realize the amount of waste that is being generated. If calculations are carried out that determine the waste generated for a given lab section, or possibly for all lab sections, it drives home the idea that waste equals money. Waste analysis can also be informative since it also allows the opportunity to teach students about incompatible chemicals and how highly reactive chemicals should be handled. As an example, strong oxidizers should not be mixed with reducing agents and organic acids should be segregated from inorganic acids.

Examples of Green Chemistry Experiments

Even though resources such as the *Journal of Chemical Education* and The University of Oregon's GEMS database are useful tools in locating green chemistry experiments for the chemical laboratory, they again are generally lacking in experiments suitable for the first-year laboratory. Because of this, some schools are making efforts to develop new experiments and redesign "old classics" to a new shade of green. St. Olaf College is one such school.

With assistance from a grant from the W. M. Keck Foundation, St. Olaf has the goal of incorporating green chemistry across the *entire* chemistry curriculum. Efforts at the first-year level have led to the redesign of four experiments with minor modifications made to an additional two. These redesigns range from replacing harmful/toxic chemicals with relatively environmentally benign options to complete overhauls of the procedures. *It is important to realize that when these procedures were redesigned, the original learning objectives remained intact.*

One such experiment that was changed involved the determination of a formula of a hydrate salt (*21*). The original experiment utilized toxic barium chloride dihydrate, Bunsen burners, and disposal of the dehydrated salt at the end of the lab. This is a classic experiment that can be carried out with a number of different salts. For example, the green chemistry twist on this experiment is the use of less toxic copper(II) chloride dihydrate, which turns from a blue green color to chocolate brown upon dehydration. Additionally, the dehydration process occurs with all student samples being placed in a drying oven set at 110 °C. These changes demonstrate the use of less toxic reagents, show increased heating energy efficiency, and minimize safety problems. To overcome the disposal issues and minimize waste, the students rehydrate the salt over a steam bath. This operation provides suitable samples of the dihydrate salt that can be reused for other lab sections. Overall, this laboratory exercise emphasizes principles of green chemistry as shown in Table I.

Table I. Green Chemistry Principles in Redesigned Hydrate Lab

Principle 1	It is better to prevent waste than to treat or clean up waste after it is formed.
Principle 3	Wherever practicable, (synthetic) methodologies should be designed to use and generate substances that possess little or no toxicity to human health and the environment.

A second experiment that was redesigned involved nickel complexation with ethylene diamine. By varying the stoichiometric amounts of ethylene diamine to nickel, students can see the different colors associated with the complexes formed. Once all of the open binding sites on the nickel are filled, the color ceases to change regardless of how much ethylene diamine ligand is added. While the lab is certainly intriguing to students because of the color changes, it poses several risks to the students. Consultation of MSDS sheets and NFPA ratings (Table II) show that nickel(II) is potentially carcinogenic. Moreover, ethylene diamine is corrosive and highly flammable and should be used in a ventilation hood. Each of these issues violates principles of green chemistry.

Table II. NFPA Chemical Ratings for Revised Stoichiometry Lab

Reagent	*NFPA Rating*	
	Health	**Flammability**
Ethylene diamine	3	3
Potassium oxalate	2	1
Sodium citrate	1	1
Iron(III) nitrate	1	0
Nickel(II) sulfate	3	0
Sodium thiocyanate	2	0

To circumvent these issues, a relatively non-toxic metal salt was chosen as starting material along with a set of non-toxic ligands: iron(III) nitrate and the ligands oxalate and citrate (Figure 2).

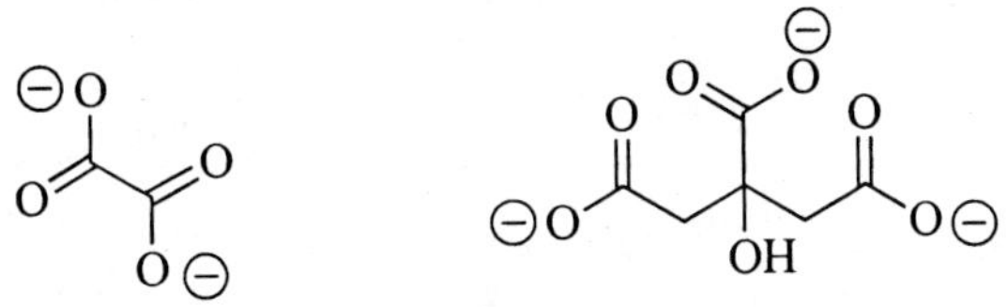

Figure 2. Structure of oxalate (left) and citrate (right) ligands.

One aspect of the original laboratory that appealed to students was the color changes. While the nickel-ethylene diamine interaction is unique in the span of colors formed, it was found in contrast that iron will form a lime-green yellow color under low ligand ratios, but then forms yellow complex solutions at higher ratios. Since the color changes are much more subtle when either oxalate or citrate is used, a simple test for open iron binding sites was needed. For this, the thiocyanate ion was chosen. Upon adding ligand at various ratios, a few drops of thiocyanate ion are added to each solution. If an open site is available for binding, the red colored complex indicative of $FeSCN^{2+}$ will form. If no sites are available for binding, the solution remains yellow. An added advantage of using the citrate and oxalate ligands is that each ligand gives its own unique metal to ligand ratio. This provides the opportunity for half of a class to work with the iron-oxalate system and the other half working with the iron-citrate system. This encourages discussion among students and leads to conversations about ligand orientation and geometric configurations of the complexes. A copy of this experiment can be obtained from either of the authors.

Many green chemistry experiments can be found in the *Journal of Chemical Education.* If the database is searched using the key word "green chemistry", many dozens of hits will be uncovered. However, only a handful of these experiments are actually geared toward the first-year curriculum. Table III gives titles and references of those experiments that directly use green chemistry as a keyword and specifically mention general chemistry students as the target audience. Since only ten percent of the retrieved hits relate to the first-year curriculum, this signifies a large window of opportunity for those involved with green chemistry curriculum development.

Another example of a green approach to first-year chemistry laboratory is the use of green analytical methods to analyze water samples taken from the field. Liz Gron at Hendrix College has pioneered this approach (*27*). This work will be discussed fully in Chapter 7 of this volume.

Table III. Examples of First-Year Green Chemistry Experiments

	Title	*Reference*
1.	Teaching Lab Report Writing through Inquiry: A Green Chemistry Stoichiometry Experiment for General Chemistry	(*22*)
2.	A Greener Approach for Measuring Colligative Properties	(*23*)
3.	Mass Spectrometry for the Masses	(*24*)
4.	Magnetic Particle Technology. A Simple Preparation of Magnetic Composites for the Adsorption of Water Contaminants	(*25*)
5.	Greening the Blue Bottle [a]	(*26*)

[a] Can also be found in the University of Oregon's GEMs Database

Student Responses to the 'Green' Revisions

As mentioned previously, St. Olaf College has made efforts to incorporate green chemistry across its entire chemistry curriculum. As a means of gauging student reaction to the greening of the first-semester general chemistry laboratory, the following short answer question was added to the end-of-semester course evaluation: *"Do you feel that you have benefited at all from doing experiments that were designed with green chemistry in mind? Please explain."* Overwhelmingly favorable responses were received. Below are representative student responses.

Student A: "Yes, it made me realize how experiments can be performed using alternate chemicals that are safe for the environment and for people."

Student B: "Yes, the fact that the chemicals were less toxic made me feel safer in lab."

Student C: "Yes, I think that they were helpful in showing that green chemistry is not just for those who are directly related in the field of chemistry in the real world, but can also be used by students."

Student D: "Emphasizing green chemistry at the undergrad level makes it more likely that the students will continue to apply it in their further work."

These student responses represent the core goals of green chemistry:

- that chemistry can be taught with environmentally friendly reagents and processes that are less harmful to students and the environment,
- that the underlying principles of green chemistry can be applied universally, and, most important, that
- when taught at early levels, the stage is set for future implementation.

Impediments to adopting green chemistry

As described in the previous discussion, green chemistry can be implemented into the curriculum by many different methods. Also, generally speaking, students see utility in learning about green chemistry and see the positive effects of its implementation. While we as educators always have the student's best interest in mind, it can sometimes be difficult to convince fellow faculty, administrators, and critics of green chemistry of its advantages. Here we will address the most common questions and concerns regarding green chemistry and how we respond to each.

"We've always done it this way"

There is always inertia toward change. A seemingly successful lecture or laboratory program already in place can offer to instructors little incentive to make alterations. The saying, "If it ain't broke, don't fix it," often applies. This means that green chemists have a real selling job ahead of them if they want to change programs already in place. Green chemistry must be not only as good as the program or individual components of a current teaching-laboratory program but even better.

It takes time to develop green chemistry experiments

We certainly agree with this statement since the development of new experiments can be a very time-consuming process. One of us (GOS) spent 20 hours per week for eight weeks during one summer just to develop a pilot laboratory program for one semester of organic laboratory. This involved developing or adapting three brand new experiments and "greening up" several others. One of the major points of this chapter and others in this book is that there are resources available to instructors wishing to adopt green chemistry into their lecture and laboratory program. Granting agencies are starting to look favorably on proposals that seek support for green chemistry implementation.

"Buy-in" by colleagues and students

Before implementing green chemistry, it is important to try to enlist the support of colleagues and administrators. Green chemistry is often quite attractive to an institution's administration because it is associated with enhanced safety and reduced waste generation. Colleagues can be the toughest sell due to the first two points covered above, but considerations of safety and waste reduction can help win them over. Most important, however, to instructors is the pedagogical value of green chemistry. If proponents can convince their colleagues that green chemistry is rigorous and is simply an alternate way of viewing key chemical concepts, implementation becomes more attractive. We have found that students buy into green chemistry readily, especially if there is strong interest on campus in improving the environment. Green chemistry often piques interest by students in chemistry who would normally have an aversion to studying the subject. The principles of green chemistry lend themselves nicely to connecting the environment with ethical perspectives throughout society.

"Greenophobia"

By "greenophobia" we mean that some who have interest in implementing green chemistry are afraid that unless they design an experiment or exercise that adheres to all the principles of green chemistry, their effort will be insufficient. We wish to emphasize that nothing could be farther from the truth. Green is a relative term, and even small incremental steps are far better than no action. Once these small steps are taken, improved, greener exercises and experiments often result because further improvements seem more obvious.

Green chemistry is not rigorous ("hippy" chemistry)

One concern that has been expressed by critics of green chemistry is that it is "soft" chemistry and not rigorous. On the contrary, green chemistry innovations must be subject to rigorous scrutiny if they are to succeed and stand the test of time. The principles of science underlying traditional chemistry are exactly the same for green chemistry.

Green chemistry doesn't teach students how to handle dangerous chemicals

We have heard such comments, and we find them interesting. If we consider the clientele we are teaching, very few of our students will go on to practice chemistry professionally. Most need to take chemistry to fill a requirement on the way to a vocational goal that has little to do with chemistry. We also argue that if green chemistry is implemented properly so that course material is presented rigorously, there is little need to be exposed, let alone know how to handle, dangerous chemicals. Later, during the more advanced stages of the career of a chemistry student, there is plenty of time to learn how to deal with hazardous materials if the need should arise. Perhaps with the students' prior training in green chemistry, they may be able to offer a safer alternative.

Conclusions

This chapter has highlighted work in green chemistry education that has occurred at St. Olaf College, but it also has described excellent efforts that have been reported at several other institutions. While examples of the implementation of green chemistry into the first-year curriculum are relatively few overall, significant work has already been done. Nevertheless, the relative paucity of work points to both the great need for further efforts and, more important, the rich, wonderful opportunities available to chemical educators to do important development and research on greening the first-year lecture and laboratory curricula. Such research and development, and the dissemination thereof, is vital for the growth of green chemistry in our education system. We must realize that as a green chemistry community, we need each other in order to get our message out.

References

1. Hileman, B. *Chem. Eng. News* **2006**, *84* (Feb 13), 70.
2. Reiner, D. M.; Curry, T. E.; De Figueiredo, M. A.; Herzog, H. J.; Ansolabehere, S. D.; Itaoka, K.; Johnsson, F.; Odenberger, M. *Environ. Sci. Technol.* **2006**, *40*, 2093.
3. Hjeresen, D. L.; Gonzales, R. *Environ. Sci. Technol.* **2002**, *36*, 102A.
4. Hjeresen, D. L.; Schutt, D. L.; Boese, J. M. *J. Chem. Educ.* **2000**, *77*, 1543.
5. Ritter, S. *Chem. Eng. News* **2007**, *85* (Feb 12), 19.
6. Ritter, S. *Chem. Eng. News* **2006**, *84* (Jul 10), 24.
7. Butters, M.; Catterick, D.; Craig, A.; Curzons, A.; Dale, D.; Gillmore, A.; Green, S. P.; Marziano, I.; Sherlock, J.; White, W. *Chem. Rev.* **2006**, *106*, 3002.
8. Tucker, J. L. *Org. Process Res. Dev.* **2006**, *10*, 315.
9. News & Announcements. *J. Chem. Educ.* **2006**, *83*, 1133.

10. Haack, J. A.; Hutchison, J. E.; Kirchhoff, M. M.; Levy, I. J. *J. Chem. Educ.* **2005**, *82*, 974.
11. *Greener Approaches to Undergraduate Chemistry Experiments*; Kirchhoff, M., Ryan, M. A., Eds.; American Chemical Society: Washington, DC, 2002.
12. Doxsee, K. M.; Hutchison, J. E. *Green Organic Chemistry: Strategies, Tools, and Laboratory Experiments*; Thomson Brooks/Cole: Belmont, CA, 2003.
13. The ACS Green Chemistry Institute. Green Chemistry Education Web Page. http://portal.acs.org/portal/PublicWebSite/greenchemistry/education/index.htm (accessed May 12, 2008).
14. *Introduction to Green Chemistry: Instructional Activities for Introductory Chemistry*; Ryan, M. A., Tinnesand, M., Eds.; American Chemical Society: Washington, DC, 2002.
15. Collins, T. *J. Chem. Educ.* **1995**, *72*, 965.
16. Green Chemistry at the University of Scranton Home Page. http://academic.scranton.edu/faculty/cannm1/greenchemistry.html (accessed May 12, 2008).
17. Song, Y.; Wang, Y.; Geng, Z. *J. Chem. Educ.* **2004**, *81*, 691.
18. Cann, M. *J. Chem. Educ.* **1999**, *76*, 1639.
19. The Greener Education Materials (GEMs) for Chemists Database. http://greenchem.uoregon.edu/gems.html (accessed May 12, 2008).
20. Green Chemistry Education Network Home Page. http://www.gcednet.org/ (accessed May 12, 2008).
21. Klingshirn, M. A.; Wyatt, A. F.; Hanson, R. M.; Spessard, G. O. Determination of the Formula of a Hydrate—A Greener Alternative. *J. Chem. Educ.*, in press.
22. Cacciator, K. L.; Sevian, H. *J. Chem. Educ.* **2006**, *83*, 1039.
23. McCarthy, S. M.; Gordon-Wylie, G. M. *J. Chem. Educ.* **2005**, *82*, 116.
24. Persinger, J. D.; Hoops, G. C.; Samide, M. J. *J. Chem. Educ.* **2004**, *81*, 1169.
25. Oliveira, L. C. A.; Rios, R. V. R. A.; Fabris, J. D.; Lago, R. M.; Sapag, K. *J. Chem. Educ.* **2004**, *81*, 248.
26. Wellman, W. E.; Noble, M. E. *J. Chem. Educ.* **2003**, *80*, 537.
27. For example, see: Nguyen, A. T.; Davenport, S. M.; Gron, L. U. Ion Chromatography for Green Environmental Analysis. *Abstr. Pap.—Am. Chem. Soc.* **2005**, *229*, INEC 86.

Chapter 6

Greening the Chemistry Lecture Curriculum: Now is the Time to Infuse Existing Mainstream Textbooks with Green Chemistry

Michael C. Cann

Chemistry Department, University of Scranton, Scranton PA 18510

It is essential that we infuse green chemistry across the curriculum from non-majors courses to majors courses. Over the last 16 years, since the beginnings of green chemistry at the EPA, green chemistry education has made significant strides but we still have a long way to go. Green chemistry educational materials have been developed, but these tend to be supplementary materials (at first outside but more frequently within existing textbooks) that are easily ignored by instructors trying to cover traditional materials in an already overcrowded course. A survey of undergraduate chemistry textbooks revealed that 33 out of 141 books contained at least some coverage of green chemistry, but the majority only mentions green chemistry once or twice in a cursory manner and generally as supplementary material. Several textbooks that "stand out in the crowd" are discussed and recommendations for improving the coverage of green chemistry in existing textbooks are given.

Sustainability and Chemistry

Many of the resources of our planet are being stretched to the limit and beyond by the burgeoning human population of the earth and advances in technology. Until recently, most advances in technology were put forth with little thought about the consequences of their environmental impact or their impact on the consumption of natural resources. If we are to foster a sustainable society then these impacts must be reduced and even eliminated. In the area of chemistry, green chemistry or sustainable chemistry has begun to answer this need.

In the United States green chemistry can trace its formal roots back to the Pollution Prevention Act of 1990 and the formation of the Design for the Environment Program at the Environmental Protection Agency (EPA) in 1991. Over the past 17 years the importance of green chemistry has become widely recognized by many companies, academics, government agencies and chemical societies. In 2005 the American Chemical Society (ACS) engaged in a study to determine how the chemistry enterprise would change over the next 10 years. This study resulted in a document, "The Chemistry Enterprise in 2015" (*1*). The following is a quote from this document:

> *"By 2015, the chemistry enterprise will be judged under a new paradigm of sustainability. Sustainable operations will become both economically and ethically essential."*

One can conclude from this statement that green chemistry will become essential to the chemical enterprise.

Currently companies are increasingly recognizing not only the economic value of green chemistry, but also the societal and environmental benefits, that is, the "triple bottom line." Sustainability has become a key driver for many chemical companies (*2*).

To prepare our students to become productive members of a sustainable society, we must bring the broader issues of sustainability and the more specific focus of green chemistry into the classroom. In 2000 Daryle Busch, who was then President of the ACS, made the following statement.

> *"Green chemistry represents the pillars that hold up our sustainable future. It is imperative to teach the value of green chemistry to tomorrow's chemists"* (*3*).

We should also add …. and to our future politicians, educators, business leaders, financiers, lawyers, economists, health professionals, and so on, for all of us need to think in a sustainable manner.

A more selfish reason for chemists to espouse green chemistry is to improve the image of the chemistry profession. For too long chemists have not paid enough attention to the environmental consequences of the compounds that they make and the procedures by which they are made. Although chemistry has brought to humankind many valuable products that help us live longer, more comfortable and productive lives, chemistry has also played a major role in bringing us environmental disasters such as Love Canal, Bhopal, Times Beach, DDT and numerous Superfund sites. It is these disasters by which we are most associated with by the public.

Greening the Chemistry Curriculum

It is now 16 years since the formal beginnings of green chemistry at EPA. So how are we doing in bringing green chemistry into the classroom? The formal focus on green chemistry education in the United States began in 1998 with a joint project of the EPA and the ACS called the EPA/ACS Green Chemistry Educational Development Project. This was developed and implemented by visionaries at the EPA and the ACS including Paul Anastas, Mary Kirchhoff, Sylvia Ware, and Tracy Williamson. The project goals included an annotated bibliography, introductory activities in green chemistry, real world cases in green chemistry, and short courses on green chemistry. Since this time ACS has been involved in producing many other green chemistry educational materials (*4*). Green chemistry teaching modules can be found on the web (*5*), and in 2001 the *Journal of Chemical Education* began specifically to encourage the submission of articles on green chemistry (*6*).

A significant question concerning the introduction of green chemistry into the classroom at the undergraduate level deals with: do we create a separate course for upper level chemistry majors dealing exclusively with green chemistry, or do we attempt to infuse green chemistry throughout the chemistry curriculum from non-majors courses to majors courses for seniors? Ideally, the answer is both. However if one has to make a choice, then the clear winner is infusion across the curriculum. Infusion across the curriculum allows all our students, from non-science majors to chemistry majors, to see how green chemistry impacts all areas of chemistry and to realize that it is not a field unto itself. It also allows all of our students (not just chemistry majors) to be indoctrinated with the issues of sustainable chemistry and sustainability.

Most of the materials that have been developed for green chemistry education are supplementary materials such as small books and websites. Instructors are encouraged to use these materials to complement the topics that they normally discuss in a particular course. We all know what happens to most supplementary materials… only the most ambitious and energetic among us even consider incorporating these resources into an already overcrowded course.

Even supplemental material within a text, such as "boxes" within or at the end of the chapter, or opening chapter vignettes are too easily and conveniently ignored in favor of the traditional material within the chapter. This often seems to be the prevailing attitude: "If it is not part of the mainstream material within the text, then it is not as important."

Green chemistry has found its way into a scattering of courses in the curriculum (*5,7*), and although the progress that has been made since 1998 has been admirable, we still have only just begun. Only when green chemistry is infused directly into the existing topics covered in mainstream textbooks will green chemistry education have "arrived". The time is now upon us to do so. With the ever-increasing emphasis on sustainable development and public opinion now beginning to favor such advances, it is incumbent that we infuse sustainable chemistry throughout our textbooks into the mainstream topics.

Infusing Green Chemistry Into Mainstream Chemistry Textbooks: An Informal Survey

In an attempt to ascertain the extent and method of incorporation of green chemistry into existing chemistry textbooks, as well as the content, we undertook two informal surveys. The first survey involved approaching the publisher's representatives during the exhibition at the 232nd ACS National Meeting and Exposition in San Francisco, September 2006. A second survey during the spring of 2007 involved contacting the publisher's representatives whose territory included the University of Scranton.

In preparation for the first survey we went to the listing of the exhibitors for the San Francisco meeting and selected the major publishers of undergraduate textbooks (*8*). We then went to each of the publisher's websites and extracted a list of undergraduate chemistry textbooks for non-majors as well as majors (*9*). In San Francisco the representatives at each of the publisher's booths were first asked, "Can you please tell me which of your textbooks have green chemistry incorporated in them?" Unfortunately several had not even heard of green chemistry. Most of the representatives were very helpful and assisted us in looking in the index and table of contents of each textbook for the term "green chemistry". Upon finding this term we would then try to ascertain the mode of insertion (*e.g.*, "box", direct infusion into the chapter topic, problem, or experiment) and the specific focus of the green chemistry.

For the second survey we first obtained the names and e-mail addresses of the publishers' representatives for the University of Scranton from the publishers' websites (*10*). We then sent each of the representatives an e-mail with the list of their textbooks (obtained from their websites) attached and asked them to indicate the following for each text:

- Is green chemistry listed in the index?
- Is green chemistry infused in the text?
- How is green chemistry inserted? Box(es) at the beginning, middle or end of the chapter? Is it infused within the chapter as part of the regular text? Are there green chemistry problems? Other? Please describe the nature of the green chemistry.

We also asked whether the publisher offered additional texts not on the list. The publishers that were involved in the first survey (at the San Francisco meeting) were:

- Benjamin Cummings
- Prentice Hall
- Houghton Mifflin
- McGraw-Hill
- W.W. Norton & Company
- Thomson
- W. H. Freeman
- Wiley
- Jones & Bartlett Publishers

All of these same publishers were contacted for the second survey; however, W.W. Norton, & Company, Wiley, and Jones & Bartlett Publishers did not complete the second survey (*11*).

As a result of the two surveys we found a total of 33 different textbooks (out of 141 textbooks) that cited green chemistry in one form or another (*12*). By publisher, Thomson led the way with twelve, followed by Prentice Hall with six, Houghton Mifflin with five, W. H. Freeman with five, McGraw-Hill with four, Wiley with one, and Jones & Bartlett Publishers, W. W. Norton & Company, and Benjamin Cummings with none. On a percentage basis (the number of books which contain green chemistry versus the number of undergraduate chemistry textbooks), Thomson was again first with 50% (12 of 24), followed by Houghton Mifflin (5 of 13) with 38%, W. H. Freeman (5 of 15) and McGraw Hill (4 of 12) with 33%, Prentice Hall (6 of 37) with 16% and Wiley (1 of 22) with 5%. These results are summarized in Table I. By level there were seven non-majors textbooks, eleven general chemistry, seven organic, five organic laboratory, one inorganic, and two environmental chemistry.

The most frequently discussed green chemistry topic was catalysts, followed by the synthesis of ibuprofen, ionic liquids, supercritical solvents, atom economy, pesticides, polymers, renewable feedstocks, and the principles of green chemistry. Other green chemistry topics that were found included

Table I. Results of Publisher Surveys

Publisher	*Number of books that cited green chemistry*	*Percentage of books that cited green chemistry*
Thomson	12	50
Prentice Hall	6	16
Houghton Mifflin	5	38
W. H. Freeman	5	33
McGraw-Hill	4	33
Wiley	1	5
Jones & Bartlett Publishers	0	0
W. W. Norton & Company	0	0
Benjamin Cummings	0	0

recycling, the Presidential Green Chemistry Challenge Awards, replacement of chromated copper arsenate (CCA) in pressure-treated wood, replacement of lead in paints, blowing agents, nanotechnology, and microwave-assisted reactions.

Textbooks that Stand Out in the Crowd

Most of the 33 textbooks that contain green chemistry only mention this topic once or twice, and most of the coverage was supplementary (*i.e.*, boxes, vignettes, or problems). However, we found a few textbooks that stand out from the rest. *Chemistry in Context* (Eubanks, Middlecamp, Pienta, Heltzel & Weaver, 5th ed., McGraw-Hill 2006) is a text for non-majors and was one of the first textbooks to incorporate green chemistry beginning with the 3rd edition in 2000. One finds an introduction to green chemistry in the first chapter, followed by seven examples of green chemistry throughout the text. The green chemistry is indicated with an icon within the text (*13*) and it can be easily found in the index under the heading "green chemistry." What also sets this book apart is the fact that the green chemistry is infused into topics that are normally covered in such a text in a seamless manner, thus green chemistry has "arrived". This serves to illustrate how green chemistry is part of the solution to many different problems. We encourage the authors of this text to expand the infusion of green chemistry into this book, and it might be helpful to the reader to indicate in the table of contents where green chemistry is found.

A second non-majors text which is exemplary is the eleventh edition of *Chemistry for Changing Times* (Hill & Kolb, Prentice Hall, 2007). In this latest edition there are 20 "Green Chemistry essays" (one for each chapter). These essays cover a wide array of topics which do a nice job of introducing many different aspects of green chemistry. The emphasis on green chemistry in this text is clearly shown by a separate table of contents of green chemistry, indication of green chemistry in the table of contents and the index, and both the back and front covers of the text "advertise" green chemistry. It was clear during our survey at the San Francisco ACS meeting that the Prentice Hall book representatives were well aware of the green chemistry in this text, and they emphasized this when they were asked about green chemistry in their textbooks. It should also be noted that the project to infuse green chemistry into this textbook was unique. The 20 essays were written by 16 individuals that were well versed in green chemistry. This project was coordinated by Kathryn Parent of the ACS Green Chemistry Institute, and represents a model for incorporation of green chemistry into other textbooks. The one major drawback of this text is that all 20 of the green chemistry essays are end of the chapter exercises. We encourage the authors to find ways to infuse these green chemistry topics into the mainstream topics of the book.

Although there are more (eleven) general chemistry textbooks that mention green chemistry than at any other level, the tendency is for only a brief mention. Perhaps the most prominent general chemistry text is the fourth edition of *Chemical Principals, the Quest for Insight* (Atkins & Jones, Freeman, 2008). The coverage of green chemistry includes a cursory introduction, three examples of infused into the text and two problems. An icon within the text indicates the green chemistry entries but what is very disappointing is the entry for green chemistry in the index points only to the introduction, and there is no indication in the Table of Contents of green chemistry.

Of the seven organic textbooks that have information on green chemistry, most have only an example or two of supplementary material on this topic. This is surprising, since many of the advances in green chemistry are associated with organic chemistry. *Organic Chemistry* (Solomons & Fryhle, 9th edition, Wiley, 2008) appears to have more coverage of green chemistry than most. In the preface of this book, readers are referred to both the ACS and EPA for more information on this subject. The Presidential Green Chemistry Challenge Awards are mentioned in chapter 1 and three examples of green chemistry are illustrated as supplementary material in other chapters of the text.

With regard to organic laboratory textbooks, one particularly stands out. As the title implies, *Green Organic Chemistry: Strategies, Tools and Laboratory Experiments* (Doxsee and Hutchison, Thomson, 2004) is dedicated entirely to green chemistry. This text grew out of a program at the University of Oregon to replace completely the existing organic laboratory curriculum with green labs. The text contains 19 examples of laboratories illustrating many traditional organic reactions and techniques, but modified or completely revised to emphasize the tenets of green chemistry.

We were pleasantly surprised to find an inorganic text with some coverage of green chemistry. *Inorganic Chemistry* (Housecroft & Sharpe, Prentice Hall, 2004) has incorporated a discussion of the Twelve Principles of Green Chemistry and four examples of green chemistry within the text.

Environmental chemistry is perhaps the most obvious field of chemistry in which to incorporate discussions of green chemistry. The eighth edition of Manahan's classic textbook *Environmental Chemistry* (CRC Press 2005) has a chapter on "Industrial Ecology and Green Chemistry." In the third edition of *Environmental Chemistry* (Baird & Cann, Freeman, 2005), the first chapter is devoted to an introduction of green chemistry and there are 14 examples of green chemistry infused into topics throughout the book, thus green chemistry has "arrived". The green chemistry topics are taken from the Presidential Green Chemistry Challenge Award winners. This book has been cited by Mary Kirchhoff, Director of the Education Division of ACS, as providing "a breakthrough in terms of integrating green chemistry into mainstream textbooks" (*14*). The fourth edition of this book is in preparation and will have additional coverage of green chemistry.

Recommendations for the Future

In order to help insure that we develop a sustainable world, it is imperative that we bring the tenets of sustainability to the people of the planet. As chemists, we can have a considerable influence on the progress that we make toward this goal by both "what we practice and what we preach". We must view all chemistry with the goal to make it greener. In order to implant this view in a student's psyche, it is necessary to infuse green chemistry across the curriculum. Educators need to look at all the materials (particularly textbooks) they use, and consider how they can infuse sustainable chemistry into the discussion.

Educators must encourage textbooks authors and publishers to integrate green chemistry into the mainstream portions of their books whenever possible. Educators need to suggest to authors ways in which green chemistry can be infused. We have found most authors to be very open to suggestions about when, where and how green chemistry can be melded into topics that are presently covered in their texts. The ACS should make this a top priority and set up teams of individuals who can aid in the infusion of green chemistry into textbooks, much the same way a team was put together for *Chemistry for Changing Times*. The ACS Exams Institute should find ways to infuse green chemistry questions into the Standardized ACS Exams. It is our understanding that green chemistry will be inserted into the next organic chemistry exam.

In order to ascertain the progress that has been made in infusing green chemistry into textbooks, a survey of these books should be undertaken on a regular basis and the results published on a website by ACS, in the *Journal of Chemical Education* and in *Chemical & Engineering News*. This website should also have a means of facilitating a discussion of how instructors have integrated green chemistry into the topics in their courses. We hope that this will encourage other instructors to do the same, and also will provide assistance to textbook authors in greening their books. The ACS, National Academy of

Sciences, EPA, and other organizations should hold workshops for authors and publishers to encourage and facilitate the infusion of green chemistry into the textbooks of the future.

References and Notes

1. For example see Baum, R. *Chem. Eng. News* **2006**, *84* (Jan 30), 3.
2. For example see the following sustainability web pages: (a) DuPont. Sustainability: A Continuing Global Challenge. http://www2.dupont.com/Sustainability/en_US/ (accessed Jun 6, 2008). (b) The Dow Chemical Company. Our Commitments: 2015 Sustainability Goals. http://www.dow.com/commitments/goals/index.htm (accessed Jun 6, 2008).
3. Henry, C. Color Me Green. *Chem. Eng. News* **2000**, *78* (Jul 10), 49-55.
4. American Chemical Society. Green Chemistry Educational Resources. http://portal.acs.org/portal/PublicWebSite/greenchemistry/education/resources/index.html (accessed Jun 6, 2008).
5. Greening Across the Chemistry Curriculum Home Page. http://academic.scranton.edu/faculty/CANNM1/dreyfusmodules.html (accessed Jun 6, 2008).
6. Journal of Chemical Education. Green Chemistry Feature Column Web Page. http://jchemed.chem.wisc.edu/AboutJCE/Features/featureDetail.php?recordID=21 (accessed Jun 6, 2008).
7. For example, see: (a) Reed, S. M.; Hutchison, J. E. *J. Chem. Educ.* **2000**, *77*, 1627. (b) Collins, T. J. *J. Chem. Educ.* **1995**, *72*, 965. (c) Cann, M. C. *J. Chem. Educ.* **1999**, *76*, 1639. (d) Cann, M. C.; Dickneider, T. A. *J. Chem. Educ.* **2004**, *81*, 977. (e) Haack, J. A.; Hutchison, J. E.; Kirchhoff, M. M.; Levy, I. J. *J. Chem. Educ.* **2005**, *82*, 974.
8. For example, see: American Chemical Society. ACS Meetings, Conferences & Expositions Web Page. http://www.acs.org/meetings (accessed Jun 6, 2008).
9. For example, see: http://www.whfreeman.com/?disc=Chemistry (accessed Jun 6, 2008).
10. For example, see: http://www.whfreeman.com/Profile/ContactUs.aspx (accessed Jun 6, 2008).
11. If the survey was not returned within five weeks, a second request was sent.
12. If the results for a particular book in the two surveys were significantly different, then we requested a copy of the book in order to reconcile the results. The most significant difference in the surveys was the confusion of environmental chemistry with green chemistry. In some instances, if we had a copy of the textbook, we made a judgment call as to whether the topic was actually green chemistry. In instances where we did not have a copy of the textbook, we generally erred on the side of accepting the judgment of the publisher.

13. At this point in time, it is important to point out the green chemistry within a text to call attention to this topic. Ideally, in the future when all chemistry is viewed with a sensitivity toward its environmental consequences, we will not have to point out that this is green chemistry.

14. Board on Chemical Sciences and Technology of the National Academies. *Sustainability in the Chemical Industry: Grand Challenges and Research Needs—A Workshop Report (2005).* http://books.nap.edu/openbook.php?record_id=11437&page=103 (accessed Jun 6, 2008).

Chapter 7

Green Analytical Chemistry: Application and Education

Liz U. Gron

Department of Chemistry, Hendrix College, Conway, AR 72032

Green chemistry seeks to reduce the hazard to the environment from chemicals and chemical processes. The most effective pollution prevention method is to avoid the use or creation of dangerous materials, rather than relegating toxins to post-processing cleanup. Despite the important role analytical chemists play in assessing environmental health, the analytical community is a relative newcomer to the field of green chemistry. Significant environmentally benign method innovations have been developed, but these are rarely described as "green". Expansion of the practice and application of green analytical chemistry will require educating our undergraduates to green principles while advancing the state of the art. Green education has made significant strides within organic chemistry, but materials for the broader undergraduate chemistry curriculum are just beginning to appear. At Hendrix College we have developed laboratories to teach green analytical chemistry using environmental samples for our introductory courses. This chapter will discuss green analytical chemistry innovations and education.

If the goal of green chemistry is to reduce the risks to humans and the environment from chemical activities *(1)*, the field of analytical chemistry provides the critical data necessary to assess human and environmental health. Therefore, it is uniquely ironic that the analytical procedures necessary to create important information on system health can produce wastes with significant negative impacts on those same systems. While the primary business of an analytical chemist is to efficiently create high quality results, it ought to be possible to do so while maintaining a protective attitude towards the environment. A wide variety of environmentally benign analytical innovations have been developed, although the term "green" is sparingly used in the literature. Environmentally benign analytical techniques have been described with a variety of labels, making them difficult to identify. In parallel to the development of environmentally benign analyses, growth of the green analytical chemistry community will also require the creation of curricular materials to educate emerging chemists. At Hendrix College, expanding the green analytical community starts with instilling a green ethic in our greenest students: those in our introductory courses. We have developed laboratory materials that teach green analytical chemistry using environmental samples for first-year students. Beyond teaching chemistry, this program will increase our students' understanding and sensitivity to the environmental consequences of their scientific choices. We expect this introduction to green ethics will continue to impact our students as they make choices in their personal and professional lives as citizens or scientists.

Why Be Green?

Green chemistry does not describe *what* we do as chemists, but rather, *how* we do it. For many of us, green chemistry draws from a passion, an underlying environmental ethic that overlays our activities. At Hendrix College, our students have been enthusiastic about our green educational emphases. However, the green chemistry community needs to be cognizant that many of our colleagues are neither instantly, nor universally, convinced of the benefits of green chemistry. It is vital that we continue to raise the consciousness of academia and industry about the power of green chemistry to improve the economic bottom line while enhancing the greater good. The practice of green chemistry unites economic success with environmental health leading to enhanced prosperity. This obliterates the traditional business model where economic and environmental growths are presumed to be mutually exclusive. An effective illustration of the connection between economics and environment is given in Figure 1 which shows the logarithmic increase in the number of health and environmental regulations during the last 130 years *(2)*. These regulations represent various significant costs, including worker training,

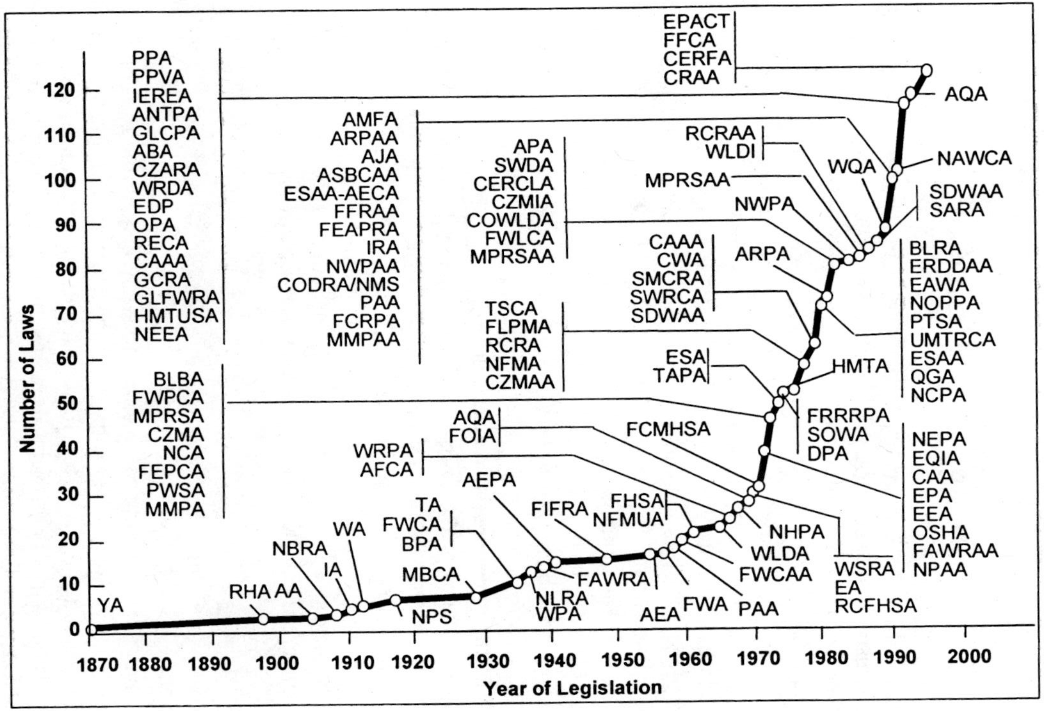

Figure 1. Graph showing the rise in environmental legislation in USA.
(Reproduced with permission from Reference 1. Copyright 1998 Oxford University).

insurance, waste remediation, compliance administration, and fines. Any businesses that can use green chemistry to reduce their exposure to, or decouple their activities from, these regulations would clearly have an economic advantage. There is a growing recognition within industry of the important economic opportunity that green chemistry presents.

Green Analytical Innovations

As illustrated by the Presidential Green Chemistry Challenge Awards *(3)*, the green chemistry movement has provided many significant developments, but analytical chemists are just beginning to focus their attention on defining "green" for their field. At this point, only a few general review articles address green analytical chemistry directly *(4-7)*. Analytical chemists uncover chemical information hidden in a sample; unfortunately, a literature search for green analytical chemistry reveals that low environmental impact procedures are also well hidden. Figure 2 gives a comparison of the publications found through an ISI Web of Science literature search for 1990 – 2006 for the keywords: "green chemistry", versus the combination of "green anal*", "clean anal*", or "green method*" *(5)*. As seen from the figures, analytical literature methods are rarely described as green although creating procedures with low environmental impact has long been a driver for method development. The lack of consistent terminology in the literature creates an intrinsic and unnecessary barrier to implementation of green analytical chemistry by making it difficult both to find the procedures and to assess the state of the field. This obstruction is particularly onerous for undergraduate educators who have neither the resources of time nor journal access to identify these innovations. A brief discussion of green analytical methods is given here with the intention of reducing the barrier to incorporating green analytical chemistry in classroom discussions.

Green Innovations: Sample Pretreatment and the Classroom

The greenness of an analytical method is largely dictated by the choice of technique in concert with the nature of the analyte and matrix. Most undergraduate instrumental courses focus exclusively on techniques (signal generation) despite the importance of pretreatment steps in real analytical schemes. Pretreatment steps can reduce data quality, consume the bulk of the analysis time, and significantly impact the environment. An important advance in green analytical education is to bring these hidden steps to light as part of the complete package that must be considered when choosing an analytical scheme.

Some methods have an intrinsic greenness advantage due to the lack of sample pretreatment. These include many classic techniques found in the undergraduate curriculum including procedures that examine aqueous inorganic

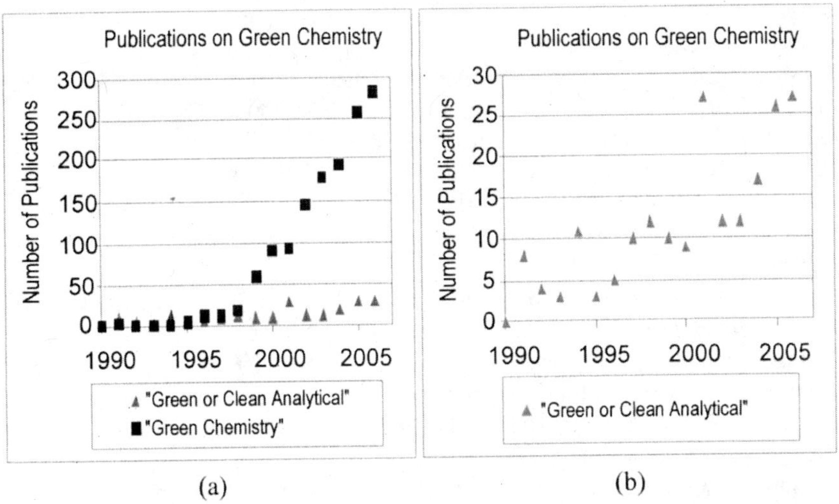

Figure 2. Number of publications resulting from an ISI Web of Science literature search for 1990 – 2006 for the keywords: "green chemistry" (■), and the combination of "green anal" or "clean anal*" or "green method*" (▲). In (a) the graph shows both keyword search results and (b) shows only the results for publications related to green/clean analytical methods. (Reproduced with permission from Reference 5. Copyright 2007 American Chemical Society).*

ions for example, pH, ion chromatography, flame and graphite furnace atomic absorption spectroscopy as well as the organic techniques of gas chromatography (GC), attenuated total reflectance spectroscopy, and total organic carbon analysis. It is a simple, but important matter to bring the greenness of these methods to the attention of our students.

Despite the exceptions listed above, most analytical schemes require substantial pretreatment that can generate large amounts of solvent and chemical waste from extraction, digestion and concentration steps. Modern analytical chemists have developed a variety of tools that minimize pretreatment wastes though few of these are discussed in modern instrumental analysis texts. Among the most common and technologically simple innovations are solid sorbants *(8)*. These materials address the thorny challenge and solvent intensive nature of organic extraction from aqueous solutions. These liquid-solid extraction formats include solid-phase extraction (SPE), and solid-phase microextraction (SPME), both of which are in use at many undergraduate institutions and are readily incorporated into the analytical curriculum.

Other new techniques for sample pretreatment may seem exotic within the present undergraduate curriculum, but these important developments allow for the acquisition of high quality data while reducing the environmental impact.

Novel sample preparation techniques include ultrasonic extractions that use high frequency acoustic waves to heat and break up samples *(9)*, as well as microwave-assisted extractions (MAE) that use long wavelength radiation for faster and less energy intensive extractions of thermally sensitive analytes *(10-13)*. Other innovations treat samples with high pressure and high temperature solvents in the liquid or in the supercritical state. These adaptations reduce the overall solvent use and speed the extractions. These methods include accelerated solvent extraction (ASE) *(14)* and supercritical fluid extraction (SFE) *(8)*.

The techniques described herein illustrate the significant progress that has been made in greening sample pretreatment by reducing or eliminating the wastes associated with analyte isolation. Although uncommon, these illustrate to our students how innovations improve the quality of the analytical results while protecting the environment. These developments educate our analytical students, and future professional chemists, to the possibilities for green method adaptations: more freedom of method choice bring with it concomitantly more responsibility to make those choices green. This knowledge will allow our students to take their places as professionals, choosing green techniques that will reduce the waste stream generated by chemical analyses.

Greenness Profiles of Environmental Procedures and NEMI

Despite the new green pretreatment methods herein, it remains difficult to readily identify green techniques, which presents a barrier to their implementation and incorporation in education. The National Environmental Methods Index (NEMI) is an important new tool that addresses this need by adding greenness assessments to its collection of over 1,000 analytical methods *(15)*. NEMI is a free, online clearinghouse containing summaries primarily for air, water and sediment analyses. The database permits searching by analyte or instrument, allowing for easy identification of major methods for a particular assay. A list of the primary method performance characteristics is also provided, including detection limits, bias, and precision. A recent ACS Green Chemistry Institute project has added greenness profiles to the method summaries in this database, now available as part of the beta test version on the website, www.nemi.gov.

The green criteria applied to the NEMI database are based in large part on government compliance regulations such as the EPA's Toxic Release Inventory (TRI) chemicals list *(16)*, and the Resource Conservation and Recovery Act (RCRA) *(17)*. A simple quartered circle symbolizes the greenness of a method (Figure 3) where each quadrant represents one of the four greenness categories: PBT (persistent, bioaccumulative, or toxic), Hazardous (chemicals listed on the TRI or RCRA's D,F, P, or U lists), Corrosive ($pH < 2$ or $pH > 12$), and Waste (> 50 g and this total waste usually includes original sample). A green quadrant

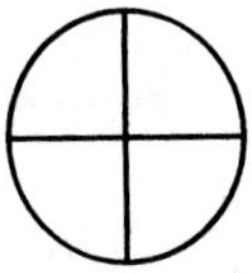

Figure 3. The greenness rating symbol from the NEMI database (Reproduced with permission from Reference 5. Copyright 2007 American Chemical Society).

indicates that the method meets that greenness criteria. The greener the circle, the more environmentally benign the method. The addition of this greenness rubric to the NEMI experimental method summaries allows the analytical chemist to quickly evaluate greenness along with the traditional analytical performance details. Method summaries are also directly linked to the source for the full method. This free, searchable collection of analytical methods promises to become an important resource for analytical students, educators and industrial chemists alike who are actively trying to enhance the greenness of their work.

Challenges of Green Educational Reform

Although modern innovations make analytical chemistry greener, national problems can rarely be solved with only action at the national level. The significant societal problem of unsustainable technology has to be solved with education that is delivered locally. The challenge of educational reform is well captured by the metaphor of eating an elephant: a project so large that it seems impossible from the start. For many educators, there is so much to accomplish that it seems a hopelessly large task from the start. To keep a project manageable, one should start where both power and support can be found. While this strategy may place unwelcome limits on creativity in the early stages of a reform, it can prevent the effort from expanding to the point where the elephant analogy crushes the forward momentum. Once the effectiveness of the approach has been illustrated, expansion can more readily take place as resources of time and money allow.

Another step necessary early in curriculum development is to understand the project goals and to note that integrating green chemistry will require two new success criteria be added to the work. The first criterion is that green educational materials must green the activities of the students and the department by reducing chemical hazards and wastes. A second, but equally important, green criterion is to overtly teach green chemical principles to the students by involving them in discussions of what green chemistry is and how to apply it to their work. The first criterion applies green chemistry directly, while the second criterion ensures that green chemistry will be applied in the future. Simply put, green chemical educators must *walk the green walk, while they talk the green talk.*

Many successful strategies exist for increasing green chemistry in the curriculum. Individual faculty members can start small by modifying existing protocols, for example, replacing obvious toxins, reducing solvent use, or adding green exercises to the pre- or post-lab reports. Many potential green program supporters have been positively swayed by modest curricular changes that illustrate the power of green chemistry to significantly reduce toxins and wastes in the undergraduate laboratory. More extensive projects generally involve adopting and adapting existing educational materials. A number of resources exist to support this type of work including the Greener Educational Materials database (GEMs) *(18)*, the ACS Green Chemistry Institute *(19)*, articles published under Topics in Green Chemistry in the *Journal of Chemical Education*, as well as the previously described NEMI database *(15)*. Finally, the most ambitious program developments generate new courses and curricula though most of these large projects started with only a few dedicated people. One excellent example of small projects that grew into large programs is described in chapter 3 of this volume. Small additions of green chemistry have been shown to have a transformational effect on students and, as students demand more opportunities for green education, on their institutions. At Hendrix College, we have used a variety of the strategies described above to green our curriculum. Our transformation to a green department began with changes in the organic program, championed by Dr. Tom Goodwin (as described in chapter 3). Since Goodwin's initial efforts, green chemistry at Hendrix College has grown to become a departmental ethic.

Green Analytical Chemistry for the Greenest Students

At Hendrix College, we have designed a new laboratory program for our greenest students, those in our introductory majors' sequence. At the beginning of the process, we selected three laboratory design criteria. We wanted the experiments to:

- be green and teach green,
- use basic analytical techniques, and
- study environmental samples within the knowledge base typical for the introductory course work.

Simple, safe, and green analytes such as nitrate, phosphate, and carbonate are environmentally relevant, compatible with the course material, and ideal for students with little previous laboratory experience. In order to give the students a more quantitative experience, we added an analytical chemistry component to provide our budding scientists with an important set of laboratory skills along with tangible experiences with gathering and assessing data. These new laboratories were given a unique name, the Green Soil and Water Analysis at Toad Suck *(20)* program (aka Green-SWAT), to provide coherence, and to reflect the unified goals.

Iron as a model toxic metal: toxic is exciting but green is safe

In two of the Green-SWAT projects, introductory chemistry students analyze for iron as a model toxic metal. Our students are fascinated by dangerous metals in the environment due to media coverage; however, it is unconscionable to use the most familiar materials, for example, Hg, Pb and As, at this level. Analyzing for iron allows us to maintain a green laboratory while engaging student interest through parallels to more toxic materials. Our two projects use modified standard molecular and atomic absorption spectroscopy protocols to analyze for iron, running for three and two weeks, respectively. These projects are similarly organized beginning with a pre-laboratory lecture and exercises, a full day for experimental work, and concluding with a laboratory notebook discussion, calculations, and data assessment. Data assessment has been an important tool for student as well as program evaluation.

Iron Analysis by UV-Vis Spectroscopy

The first project in the iron series occurs at the end of the first semester and analyzes for iron by UV-Vis spectroscopy. The pre-lab work introduces the environmental, analytical, and green chemistry aspects of the project as well as the details of the experiment. We use simple primers to link the new topics to the knowledge base typical of introductory students. To add green chemistry to the project, we have created an online exercise employing the Presidential Green Chemistry Challenge Awards. This exercise helps students make the connection between our green analyte, iron, and more environmentally exciting, but hazardous, metals. We introduce the students to the environmental problems associated with pressure-treated landscape timbers using chromated copper arsenate (CSA) as a preservative. We ask the students to use the web to research the health and environmental risks as well as find an alternative procedure. Inevitably they find their way to the 2002 Presidential Green Chemistry Challenge Award given to Chemical Specialties, Inc (now owned by Viance) for the development of an environmentally benign wood preservative using alkaline copper quaternary (ACQ). Replacement of the CSA preservative with the ACQ is expected to reduce the industrial use of arsenic in the United States by about 90% *(3)*. The Presidential Green Chemistry Challenge Awards are a wonderful resource for readily adding green educational components to existing programs. In this case, the chemical concepts are simple while the chemistry is important. These awards illustrate the positive power of chemistry to our students, and show our students that chemists are among the good guys, developing the solutions to pollution.

In the laboratory portion of the project, the students quantify iron in real and artificial surface water samples by UV-Vis spectroscopy. The iron is complexed to the *o*-phenanthroline (phen) ligands in a buffered solution to create a highly colored orange complex, $[Fe(phen)_3]^{+2}$. The intensity of the complex color is proportional to the concentration, following Beer's Law. Students create a standard series and prepare a surface water sample using modified standard protocols *(21)*. We use autodispensers to dispense corrosive reagents and provide the stock iron solutions; this equipment reduces exposure and ensures that the experimental work can fit within the three hour laboratory period. Students measure of the absorbance of their standard series as well as their surface water sample on a spectrometer at λ = 508 nm. Students complete the experimental write-up, calculations, data analysis, and assessment during the subsequent laboratory period.

Iron Analysis by FAA Spectroscopy

The second iron project in the Green-SWAT program analyzes for iron by flame atomic absorption (FAA) spectroscopy in surface water samples and comes at the beginning of the second semester of the introductory chemistry course. This experiment also uses standard protocols *(22)* with minor adaptations similar to those used in the UV-Vis experiment. Students find the FAA project inherently easier than the related UV-Vis experiment due to their previous experience with spectroscopy and the simplified solution preparation. This allows the project to be completed in two laboratory periods rather than the three required for the UV-Vis analysis. Since the project is less complicated, we have added an extra challenge. Along with their standard solutions for their calibration curves and surface water samples, students prepare a standard reference material to assess accuracy of their results. This graded assessment has a tendency to unnerve our students initially, but it gives them a feeling of accountability for the quality of their results. Students present their surface water sample, reference material, and standard solutions to a laboratory assistant who runs the FAA instrument, but students are required to record their own data. Calculations, data analysis, and assessment are usually accomplished on the same day as the experimental work.

Happy Students and Assessment Tools

It is important for the integrity of the green chemistry movement that we prove that new green materials are as effective at attaining student learning goals as the traditional materials they replace. Assessment has been a valuable aspect of our program development. Students generally enjoy green chemistry laboratories, and that can give rise to concerns about program quality. Surely, if students are enjoying a laboratory, it must be watered down. Effective assessments prove to local and national stakeholders that students can enjoy learning chemistry, while participating in a intellectually rigorous program.

We have used a number of tools to monitor the effectiveness of our program: student precision and accuracy data from the iron projects, a Student Assessment of Learning Gains (SALG) *(23)*, a laboratory practical, as well as a written final exam. These tools are listed against our student learning goals as seen in Table I. All these tools are used every year although they are adapted slightly in order to focus on different learning goals.

Table I. Student Learning Goals and the Assessment Tools for the Green-SWAT Program

Learning Goals & Assessment Tools	*Accuracy & Precision Data*	*SALG* [a]	*Lab Practical*	*Written Exam*
Quantitatively transfer solids & liquids	√	√	√	√
Create analytical solutions (volumetric flasks, pipets and burets)	√	√	√	√
Use a Spec20 (with basic instructions)	√		√	
Use EXCEL manage numbers and graph	√	√		
Use calibrations curves	√	√		√
Use basic statistics		√		√
Understand spectroscopy		√		√
Know environmental action of ions		√		√
Define and explain green chemistry		√		√

[a] Student Assessment of Learning Gains; SALG, is adapted from original work by Elaine Seymour *(23)*

Collecting this assessment data is time consuming, but we need to assure ourselves and our colleagues of the efficacy of this program in teaching our traditional goals of quantitative skills as well as the new program goals of green and environmental chemistry. To evaluate the Green-SWAT program, we tracked student precision in both iron projects and student accuracy from the standard reference material used in the FAA spectroscopy laboratory. Student precision in both experiments was measured as the % relative error of the slope (e_m/m %) of the student generated calibration curve where the error of the slope, e_m, was calculated by the LINEST function in Microsoft EXCEL and m was the slope of the calibration curve. The % relative error in the slope exaggerates the differences between the data. Student accuracy was evaluated as the % error between the % iron in the standard reference material and the % iron experimentally derived by the students.

We were gratified by the excellent precision that student were able to attain. By the FAA spectroscopy experiment, over 90% of the students had less than five per cent error on their calibration curves. However, we were disappointed and puzzled by the poor results these same students had in their accuracy. Only 40% of the students were within the target of < 10% error on their reference material and over 34% of the students had over 20% error. We suspected that accuracy errors stemmed from the quantitative transfer of a solid and a solution, the only skill unique to the accuracy work. We confirmed our suspicions by designing a laboratory practical to investigate quantitative analysis skills.

After much complaining about our inattentive students, we turned to strengthen the quantitative transfer directions in our materials. Much to our embarrassment, we found that quantitative transfer had never been intentionally targeted as a laboratory skill. It had been mentioned in passing, discussed in the pre-lab directions, but it had never been written into the laboratory directions. A redesign of our program materials in 2006-2007 showed that we had effectively addressed this problem. The students' precision skills remained intact while the number of students attaining our accuracy goal of < 10% error on the FAA analysis had jumped from 39% to 69%. Additionally, the number of students considered outside of the acceptable range of < 20% error on accuracy had dropped from 34% to 13%. For our project, assessment has been a critical development tool, demonstrating the strengths as well as identifying weaknesses, of our Green-SWAT program materials.

A Final Initial Analysis of Green Analytical Chemical Education

It is still in the early days for green analytical chemistry education. Analytical chemists have made significant innovations in greening procedures, giving analytical chemists more choices of greener methods, which are slowly making their way into the undergraduate curriculum. Collections such as the NEMI database make these methods more readily identified giving analytical educators and chemists the opportunity, and responsibility, to choose low environmental impact procedures. The Green-SWAT program at Hendrix College has been demonstrated to be effective for teaching introductory students practical analytical laboratory skills using green procedures on environmental samples. Our emphasis on analytical analysis of environmental samples has been an ideal place to emphasize green chemistry since it works to break down the apparent paradox between doing environmental analyses and caring for the environment. Although an introductory program cannot create experts, our program works to create environmentally and scientifically "savvy" students through analytical chemistry. Our students are able to ask the important questions about data, methods, and environmental impact. These green laboratories will have little impact on our local waste stream, but our green-educated students will go on and make choices in their professional and personal lives informed by their green ethic, which will have a significant and positive environment impact.

References

1. Anastas, P. T.; Warner, J. C. *Green Chemistry: Theory and Practice*; Oxford University Press: Oxford, U.K., 1998.
2. *Green Chemistry: Frontiers in Benign Chemical Synthesis and Process*; Anastas, P. T., Williamson, T. C., Eds.; Oxford University Press: Oxford, U.K., 1998; p 4.
3. The Presidential Green Chemistry Challenge: Award Recipients 1996–2008. U.S. Environmental Protection Agency Green Chemistry Web Site. http://www.epa.gov/gcc/pubs/docs/award_recipients_1996_2008.pdf (accessed Jun 12, 2008), EPA document 744F08008, June 2008.
4. Anastas, P. T. *Crit. Rev. Anal. Chem.* **1999**, *29*, 167.
5. Keith, L. H.; Gron, L. U.; Young, J. L. *Chem. Rev.* **2007**, *107*, 2695.
6. de la Guardia, M.; Ruzicka, J. *Analyst (Cambridge, U.K.)* **1995**, *120*, 17N.
7. Namiesnik, J. *J. Sep. Sci.* **2001**, *24*, 151.
8. Wrobel, K.; Kannamkumarath, S.; Wrobel, K.; Caruso, J. A. *Green Chem.* **2003**, *5*, 250.
9. Barriada-Pereira, M.; Gonzalez-Castro, M. J.; Muniategui-Lorenzo, S.; Lopez-Mahia, P.; Prada-Rodriguez, D.; Fernandez-Fernandez, E. *Talanta* **2007**, *71*, 1345.
10. Zuo, Y. G.; Zhang, L. L.; Wu, J. P.; Fritz, J. W.; Medeiros, S.; Rego, C. *Anal. Chim. Acta* **2004**, *526*, 35.
11. De Orsi, D.; Gagliardi, L.; Porra, R.; Berri, S.; Chimenti, P.; Granese, A.; Carpani, I.; Tonelli, D. *Anal. Chim. Acta* **2006**, *555*, 238.

12. Cava-Montesinos, P.; Rodenas-Torralba, E.; Morales-Rubio, A.; Cervera, M. L.; de la Guardia, M. *Anal. Chim. Acta* **2004**, *206*, 145.
13. Lucchesi, M. E.; Chemat, F.; Smadja, J. *J. Chromatogr., A* **2004**, *1043*, 323.
14. Robinson, J. W.; Frame, E. M. S.; Frame, G. M. I. *Undergraduate Instrumental Analysis*, 6th ed.; Marcel Dekker: New York, 2005.
15. National Environmental Methods Index. http://www.nemi.gov (accessed Jun 12, 2008).
16. Emergency Planning and Community Right-to-Know Act, Section 313, Toxics Release Inventory (TRI); the most recent chemical list available in 2008 is for the reporting year 2006 See: U.S. Environmental Protection Agency. Toxics Release Inventory (TRI) Program TRI Chemicals Web Page. http://www.epa.gov/tri/chemical/ (accessed Jun 12, 2008).
17. Identification and Listing of Hazardous Waste. *Code of Federal Regulations*, Part 261, Title 40; http://ecfr.gpoaccess.gov (accessed Jun 12, 2008).
18. The Greener Education Materials (GEMs) for Chemists Database. http://greenchem.uoregon.edu/gems.html (accessed Jun 12, 2008).
19. The ACS Green Chemistry Institute. Green Chemistry Education Resources Web Page. http://portal.acs.org/portal/PublicWebSite/greenchemistry/education/resources/index.htm (accessed Jun 12, 2008).
20. Toad Suck is the colorful and historically correct name for the Arkansas River crossing closest to Conway, AR.
21. AOAC methods. Iron in plants, 937.03, and Iron in Flour, 944.02. In *Official Methods of Analysis of the AOAC International*; Horwitz, W., Ed.; Vol. 17th edition. AOAC International: Gaithersburg, MD, 2003.
22. AOAC method. Cadmium, chromium, copper, iron, lead, magnesium, manganese, silver, and zinc analysis in water, 974.27. In *Official Methods of Analysis of the AOAC International*; Horwitz, W., Ed.; Vol. 17th edition. AOAC International: Gaithersburg, MD, 2003.
23. Student Assessment of Learning Gains (SALG) is adapted from original work by Elaine Seymour: http://www.wcer.wisc.edu/salgains/instructor/SALGains.asp (accessed Jun 12, 2008).

Chapter 8

Linking Hazard Reduction to Molecular Design

Teaching Green Chemical Design

Nicholas D. Anastas[1] and John C. Warner[2]

[1]Principal, Poseidon's Trident, LLC, Milton, MA 02186
[2]President and Chief Technology Officer, Warner Babcock Institute for Green Chemistry, Woburn, MA 01801

Green chemistry, defined as the design of chemical products and processes that reduce or eliminate the use and generation of hazardous substances, is making its way into all aspects of the chemical enterprise as well as into the middle school, high school and university chemistry curriculum. An important component of green chemistry is minimizing toxicity and other hazards as part of the chemical design phase. Often missing from the instruction of chemists, however, is the connection between molecular structure and hazard. Students and practicing synthetic chemists need to be aware that the hazardous nature of a substance can be controlled through structure manipulation. Through careful molecular design, chemists can develop new substances that maintain functionality while minimizing hazard. This chapter outlines the basic components that must be included in an approach that links hazard reduction to molecular design as part of a comprehensive and systematic approach to green chemical design.

Introduction

The emergence of green chemistry in research and in industrial applications makes it clear that the long-term basis for achieving the full potential of green chemistry will be found in education. It is only by training future chemists and molecular designers in the principles of hazard and hazard evaluation that chemistry as a discipline will have the perspective and the framework necessary to design products, processes, methodologies, and techniques that minimize hazards before chemicals are synthesized.

Bringing green chemistry to the classroom has been quite successful. A number of papers have been written on various aspects of introducing green chemistry into the curriculum (*1-4*). There is a recognized need that green chemistry must be included at all levels of education, from the first introduction to science in middle school and high school through university training and beyond to practicing chemists in academia and industry. Pfizer Pharmaceuticals has developed "Recipe for Sustainable Science: An Introduction to Green Chemistry in the Middle Schools" (*5*). The University of Scranton has developed web-based materials that provide educators with green chemistry modules to use in their classrooms. Greener Education Materials for Chemists or GEMs, is an interactive, web-based resource providing educational material on green chemistry that encourages individuals to contribute instructive resources (6) The success of teaching green chemistry can be used as an inspiration and a challenge to begin similar efforts in the introduction of green chemical design into education.

The molecular basis of hazard is primarily concerned with the design of safer chemicals or the manipulation of molecular structure for minimized hazard (*7*). Perhaps because of this topic's close link to toxicology, many instructors in the field of traditional chemistry are uncomfortable incorporating the basic principles into their existing chemistry courses for fear of expending valuable class time attempting to turn every chemist into a toxicologist. This fear is unjustified. There are many places in the existing curriculum where approaches to designing safer chemicals can and should be presented, as well as opportunities for developing special independent courses on the subject. Examples will be presented in this chapter that cover the range of hazards that should be included in a complete treatment of designing safer chemicals. For example, the effects of electrophilic or nucleophilic substitution can be discussed in terms of their adverse effects on alkylation of biologically important molecules (*e.g.*, DNA, RNA, proteins). Thermodynamic principles associated with explosives can be worked into existing lecture material in physical chemistry lectures. While the level of detail presented will vary with the focus of the course and with the training of the instructor, the fundamental framework of the topic will stay the same.

The Target Audience

The primary audience for this teaching approach includes university instructors, students and practicing synthetic chemists, all of whom should become familiar with the fundamental tools that can be used to characterize hazard as an integral component of designing safer molecules. Chemists must intelligently and rationally approach chemical design. This requires an understanding of the principles of hazard and training to identify potential danger in a molecule.

A secondary audience includes toxicologists and allied environmental scientists. In the same way that not all chemists need to be trained as toxicologists, not all toxicologists need to be trained as chemists. However, cross-training is an integral part of the success of green chemical design. The more appreciation a toxicologist has for the thought process a synthetic organic chemist goes through when designing a chemical, the more likely it is that information shared between chemists and toxicologists will help to mitigate or eliminate hazard. This may give rise to a new type of investigator, perhaps a "green toxicologist" whose primary objective is to look at hazards with the specific goal of identifying molecular features that can be manipulated as a critical component of developing and articulating the structure-design relationship.

Each of the elements listed below is essential to teaching green chemical design:

- The Molecular Basis of Hazard (the structure–hazard link)
- Types of Hazard
  - Physical
  - Toxicological
  - Global
- A framework for green chemical design (*8*)
  - Influencing the mechanism/mode of action
  - Addressing functional groups
  - Developing and applying structure-activity relationships (SAR)
  - Designing for reduced hazard using kinetics and dynamics
  - Influencing availability

The Molecular Basis of Hazard

A core tenet within green chemistry is that hazard is another chemical property just like boiling point, solubility, or color (*9*). Familiarity with the relationship between hazard reduction and molecular design, and identifying the opportunities that exist for designing safer chemicals, must be core components of the education of all chemists as well as toxicologists and environmental scientists.

Designing chemicals that are less toxic is one of the principles of green chemistry and the focus of this chapter is to present an expansion of this principle into a comprehensive approach for designing safer chemicals that can be used in the classroom. The same tools of chemistry used to examine structure, kinetics, dynamics and mechanisms that have revolutionized the electronics, transportation, and communication industries can also be used to design minimal hazard into chemical products. Minimized hazard needs to be understood as a performance criterion in the design of safer chemicals. ***The existence of hazard should be recognized as a design flaw.***

Through appropriate illustrations, students can understand that intrinsic chemical hazard is not something to accept as part of chemistry, but a property to be controlled through rational, informed design as part of the responsible practice of chemistry. There are numerous examples that can be gathered from the pharmacology and toxicology literature that can provide examples of how understanding the fundamental chemical principles associated with the manifestation of therapeutics and biochemistry can be used to rationally design safer chemicals (*10*).

Types of Hazard

Hazard is a broad term defined as the ability of a substance to result in consequences that are adverse to human health and the environment and encompasses several subcategories (Table I). The first time students enter a laboratory they are cautioned about *physical* hazards such as flammability, explosivity, and corrosivity. Physical hazards can be defined as events that cause injury or significant disruption at a well defined, localized level. A discussion of how these physical hazards are derived from their molecular structure is an important component in making students aware that they can influence the potential for hazard as they design new molecules.

This same principle holds for discussions of *toxicological* hazards. Often, students are only exposed to the fact that something ***is*** toxic rather than engage in a discussion of the principles of ***why*** it is toxic. Addressing the subject of toxicity in a manner that describes the principles of toxicity and illustrates the ability of a synthetic chemist to control toxicity through molecular design will again emphasize the intimate connection between molecular structure and hazard. Seldom is hazard a desired property of molecular structure; however, in those cases where it is desirable (*e.g.*, pesticides, antibiotics, anticancer drugs and the focused application of explosives), those hazards need to be as precisely targeted (selective) and only as enduring as absolutely necessary. Selective

Table I. Hazard Categories and Selected Examples for Designing Safer Chemicals

Hazard Category	*Selected examples*
Physical	Corrosivity Explosivity Extremes of pH Flammability Strong oxidizers or reducers Radioactivity
Toxicological	*Human, Non-organ directed* Cancer Genetic alterations Developmental deficiencies *Human, Organ Directed* Blood, Immune system Liver, Kidney, Nervous System Heart and vascular Reproductive and endocrine Eye and Skin *Aquatic* Lethality Decreased Fecundity *Terrestrial Wildlife* Lethality Endocrine disruption *Plants* Lethality Decreased growth
Global	Global Climate Change Ozone depletion

toxicity is defined as the ability to affect one type of cell or organisms without affecting another (*11*). The desired or economic species remains unaffected while the undesired or uneconomic species is injured in some way.

Global hazard examines adverse consequences that occur on a larger scale area and include stratospheric ozone depletion and effects of chemicals on global climate change.

Risk is defined here as the probability of an adverse consequence occurring through consideration of a number of exposure and hazard related parameters often described by the following relationship:

$$\text{Risk} = \text{Hazard} \times \text{Exposure}$$

Risk reduction can be accomplished by reducing hazard, exposure or both. Synthetic chemists tasked with designing safer compounds focus on hazard reduction through molecular design.

Approaches to Green Chemical Design

Influencing the mechanism or mode of action

Over the course of the past generation or more, there has been a great deal of research on the mechanisms of action detailing how chemical substances manifest hazard, whether physical, toxicological, or global. Though we do not have this level of understanding for most chemicals, it is the most powerful tool that students and practitioners can use in the design of safer chemicals. By understanding the reactions and interactions that are essential in elucidating the molecular sequence of events that lead to an undesirable endpoint, chemists can design structures such that these reactions either cannot take place or are greatly disfavored, thereby reducing the intrinsic hazard.

The current trend in hazard assessment is toward understanding the molecular mechanisms associated with toxicity, as well as more fully characterizing mechanistically the hazards associated with physical phenomena, for example ozone depletion and global climate change. This movement toward molecular hazard assessment can complement the philosophical approach used in teaching green chemical design to students by clearly linking physical, toxicological or global hazard with fundamental chemical structure. Chemical toxicologists will be important contributors to a multidisciplinary team of scientists that have as their goal the design of useful products that are less hazardous.

Epoxidation of benzene

As part of the discussion of aromatic ring epoxidations common in undergraduate organic chemistry classes, the fact that benzene must be metabolized enzymatically to the epoxide to act as the ultimate carcinogen is a good illustrative example of the concept of bioactivation. It can be slipped nicely into an existing organic chemistry lecture niche. An instructor can develop a design challenge to the class to describe the conditions that favor or disfavor the formation of an aromatic epoxide and provide an example of a chemical structure that is likely less hazardous. An example may include the polyaromatic hydrocarbons (PAH).

Redesign of Felbamate

Felbamate is an antiepileptic drug that was approved for clinical use by the Food and Drug Administration in 1993 for those patients that were unresponsive to traditional antisiezure medications (*12*). Within a year of its introduction to the market, patients began to experience severe liver toxicity and aplastic anemia prompting FDA to issue a "black box" warning identifying the extreme toxicity. Mechanistic toxicology studies identified 2-phenylpropenal as the active metabolite responsible for the adverse response.

Medicinal chemists redesigned felbamate so that 2-phenylpropenal could not be formed as a metabolite. The redesigned compound is fluorofelbamate. All three structures are presented below (Figure 1). An instructor could present this redesign as a case study asking students to describe the fundamental chemistry behind the successful molecular design. What feature of adding a fluorine molecule in place of the hydrogen reduces or eliminates the formation of the active metabolite? Can the student propose any other molecular changes to achieve the same endpoint?

Antibiotics

Molecular design of antibiotics has resulted in improved therapies for combating infectious diseases. Cephalosporins are related to penicillins in that both classes require the β-lactam functionality to inhibit bacterial cell wall growth. An example of molecular design for decreased toxicity is the development of an effective yet less toxic member of the cephalosporin antibiotics. The structure of cephaloridine (Figure 2) shows a cationic nitrogen in the terminal heterocycle. Another cephalosporin, cephalothin (Figure 3), lacks this feature and possesses lower kidney toxicity (*13*). The positively charged cephaloridine is taken into the cell by a trans-membrane carrier protein but because the molecule remains charged once inside the kidney cells, it cannot

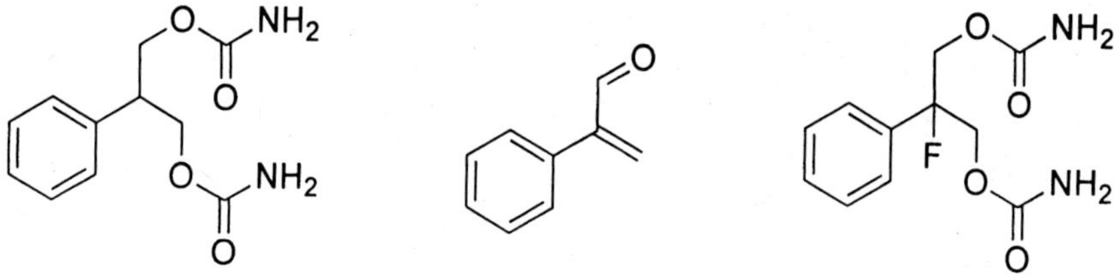

Figure 1. Structures of Felbamate, 2-phenylpropenal and Fluorofelbamate.

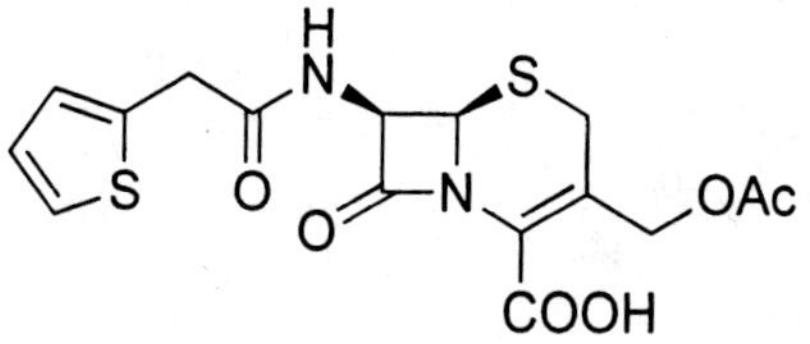

Figure 2. Structure of Cephalothin.

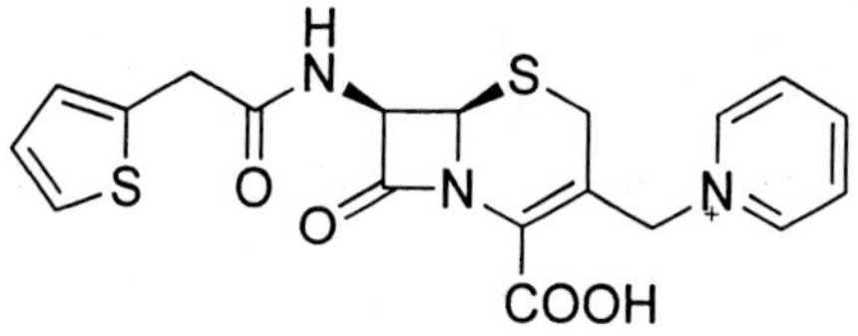

Figure 3. Structure of Cephaloridine.

cross back through the membrane and therefore accumulates and leads to kidney damage.

The influence of structure on the acute toxicity of nitriles can be used in a lecture to illustrate the fundamental structure–hazard link. Many chemists who have participated in an organic chemistry laboratory section will remember the warning that the aroma of burnt almonds may indicate that cyanide has been released. An opportunity exists for the laboratory instructor to explain where and how cyanide evolved and that it is an acutely dangerous chemical. . This example that can be incorporated into the first-day lab overview is to include a discussion of the source and toxicity of nitriles to humans. Nitriles are ubiquitous in a synthetic laboratory and can liberate cyanide under certain circumstance. The structural features that are associated with a greater chance of cyanide being liberated from a nitrile have been studied and reported (*14*).

One mechanism of nitrile toxicity requires that the cyano moiety be released from the molecule (Figure 4). This particular example illustrates the *in vivo* transformation by cytochrome P450, a very common oxidation enzyme found primarily in the liver but also in other tissues.

Studies have concluded that the rate of abstraction of the hydrogen atom next to the cyano group will determine the potency of the nitrile (*15*). Armed with just this information, students can be asked to identify the structural features that influence the rate of hydrogen abstraction. The factors that influence the rate of cyanide release are the same as those familiar to organic chemistry students, namely electronic and steric effects. A compilation of these structural features can be assembled for other classes of molecules and be apllied by the students to other molecules.

Molecular scientists can continue to have a profound effect on global hazards. As the topics of climate change, stratospheric ozone depletion, and the depletion of other resources are introduced, demonstrates that these are not merely problems of bad practices, behavior, or circumstances; rather, they are issues that can be greatly affected by appropriate and enlightened molecular design by chemists. Discussing classical studies of the destruction of stratospheric ozone by chlorofluorocarbons (CFC) provide instructors with an opportunity to describe how the high bond strength of the carbon-halogen bond imparts the molecular stability that is necessary to reach their target, in this case, the troposphere. Replacing halogen atoms on short chain hydrocarbons with hydrogen atoms, reduces the stability in the atmosphere decreases thereby decreasing hazard, in this case ozone destruction.

Addressing functional groups

In cases where detailed mechanisms of action do not exist, there is often an understanding that the presence of certain functional groups may result in a hazardous endpoint. Where the possibility exists to simply remove these

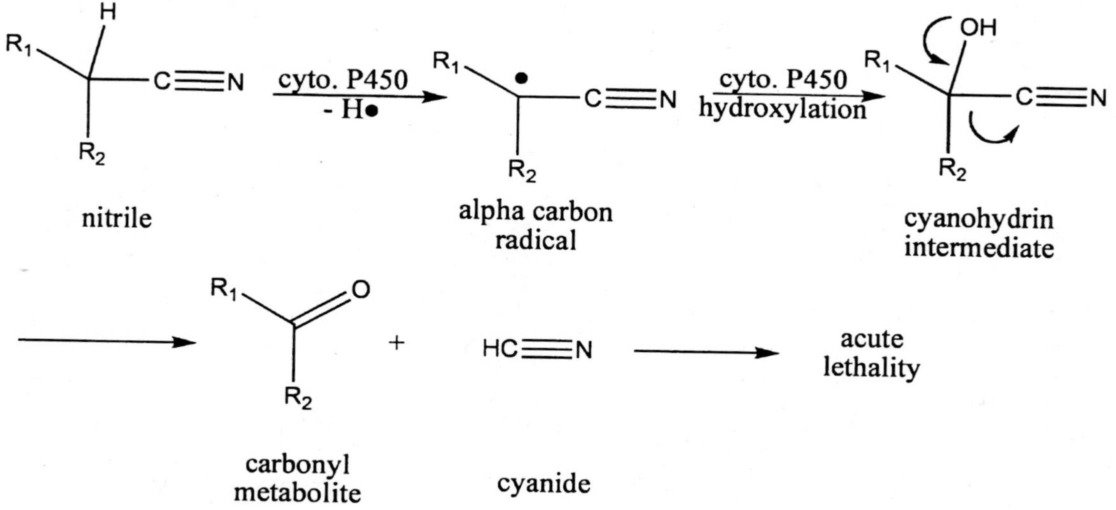

Figure 4. Mechanism of cyanide release from nitriles.

functional groups without sacrificing the desired purpose or function of the molecule, it is possible to greatly reduce or eliminate the ability of the molecule to manifest this particular hazard. As part of a discussion on functional groups important in organic reactions, examples of hazardous functional groups can be included to demonstrate this concept.

Explosives and molecular structure

Certain structural arrangements of molecules containing nitrogen and oxygen are more closely associated with the potential to explode than others. The presence of certain functional groups, for example peroxides, increase the likelihood of explosions. After an introduction of the principles behind explosives chemistry, instructors can present a series of unlabeled structures and ask the students to predict which of the structures is likely to possess the features of an explosive. Some examples can be trinitrotoluene (TNT) and Her Majesty's Explosives (HMX) and picric acid (Figure 5a) and some non-explosive structures (Figure 5b) that contain nitrogen and oxygen, but not in the correct locations to be explosive. The learning point of the exercise is to illustrate the necessary structural features of oxygen and nitrogen in the correct configuration.

Structure-activity relationships

The relationship between a chemical's structure and its biological action has been studied extensively for over a century (*16*). In cases where there is not a complete understanding of the mechanism/mode of action or where the influence of functional groups is not known or obvious, there is a vast body of knowledge on how different structural features within a class of chemicals may correlate with various levels of hazard. Structure–activity relations (SAR) or their mathematical treatment, Quantitative SAR (QSAR) have been developed for myriad endpoints including cancer, developmental and reproductive effects, aquatic toxicity, boiling points, water solubility and many others hazard endpoints. An instructor therefore has many opportunities to incorporate the concept of SAR at several points in the curriculum.

Aromatic amines and SAR

Exposure to certain aromatic amines, some of which are considered potent mutagens, is associated with an increased risk of cancer. N-nitroso- or aromatic amine functional groups serve as structural alerts and are used to prioritize new chemicals for greater study under the Toxic Substances Control Act. Variations on the location and number of amine groups as well as the number of aromatic

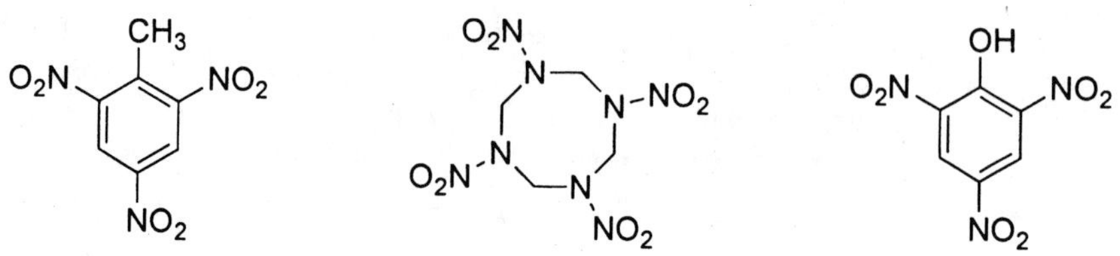

Figure 5a. Structures of nitrogen- and oxygen-containing explosives: trinitrotoluene (TNT), Her Majesty's Explosive (HMX) and picric acid.

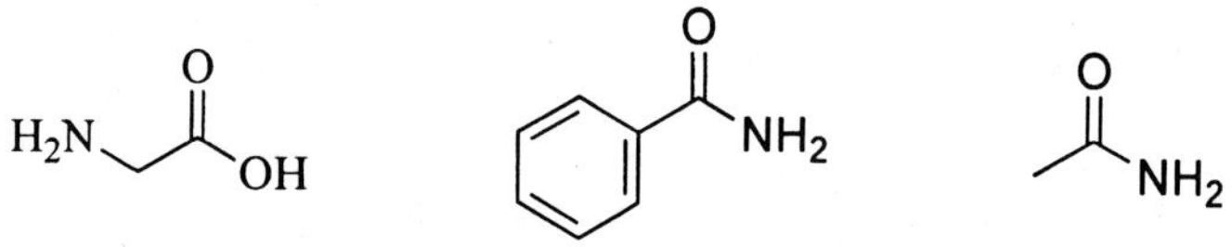

Figure 5b. Structures of nitrogen- and oxygen-containing non-explosives: glycine, benzamide, acetamide.

rings can be used to demonstrate how a simple SAR is developed by using structure and toxicity. Table II can be used as the demonstration exercise to introduce students to the relationship between structure and toxicity.

Using this type of approach, where actual toxicity data are matched against structures and conclusions are drawn based on inspection, is valuable for the students as a stepping- stone for the more advanced computer-based applications associated with QSAR.

Kinetics and Dynamics

The fundamental principle of toxicology is the concept that the sixteenth century physician Paracelsus articulated in the 1500s: *sola dosis facit venenum* or "the dose makes the poison". The modern version of this observation is the dose-response relationship, which is experimentally and theoretically supported through pharmacokinetic and pharmacodynamic experimentation. Pharmacokinetics is concerned with the study of the time course of the disposition of drugs, specifically absorption, distribution, metabolism and elimination, often referred to as ADME. In non-technical terms it can be thought of as what the body does to the chemical. An understanding of the pharmacokinetic (in the case of drugs) or toxicokinetic (all chemicals) profile is critical to estimate the concentration at the site of action. The process of drug discovery and development relies heavily on the results of ADME data, both empirical and modeled, to modify the structure of lead compounds to improve efficacy, reduce toxicity and bring them to market.

Absorption

Absorption is the process whereby a chemical crosses biological membranes into the general circulation. Many factors can influence absorption including molecular size and charge, as well as surface area, blood flow and concentration at the exposure site. For the purposes of showing synthetic chemists how structure can influence absorption, differences in ionization can be used as an example of one influence on absorption. The fact that neutral chemicals are absorbed more efficiently than charged species can be introduced when students are exposed to the concept of pH. The outside of a biological membrane contains proteins, sugars and phosphate groups and is generally charged. The interior of most biological membranes contains hydrocarbons of various structures and is uncharged.

For example, in an exercise one might have students calculate the percent of salicylic acid (active molecule of aspirin) versus its anionic form at pH 1.2 and pH 7.4. These values correspond to the average gastric pH and the pH of the plasma respectively. This can be done using the Henderson-Hasselbach equation:

$$\text{pH} = \text{pKa} + \log\,(\text{ionized species} \div \text{unionized species})$$

Given that the pKa of salicylic acid is 3.0, the solution is:

pH 7.4

$$\log\,(\text{ionized species} \div \text{unionized species}) =$$
$$\text{pH} - \text{pKa} = 7.4 - 3.0 = 4.4$$
$$\text{antilog}\,(4.4) = 2.51 \times 10^{4}$$

pH 1.2

$$\log\,(\text{ionized species} \div \text{unionized species}) =$$
$$\text{pH} - \text{pKa} = 1.2 - 3 = -1.8$$
$$\text{antilog}\,(-1.8) = 1.58 \times 10^{-2}$$

This illustrates that salicylic acid is well absorbed from the stomach because it exists primarily in the neutral state and less well absorbed in the parts of the digestive system where the pH is closer to 7.4.

Table II. SAR for Aromatic Amines

Compound/Structure	*Endpoint: Carcinogenicity*	*Rational for Toxicity*
Benzidine 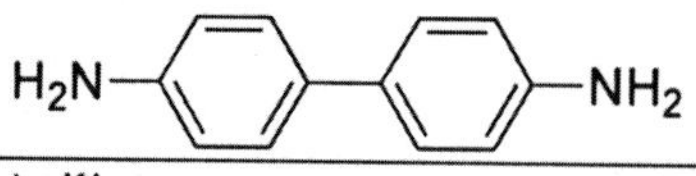	Extremely toxic; Carcinogenic	Amine groups occupy both 4 and 4' positions
Aniline 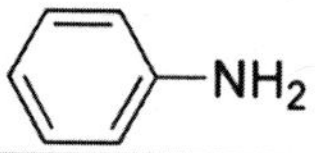	Not carcinogenic	Lacks a second aromatic ring
2,4'-Biphenyldiamine 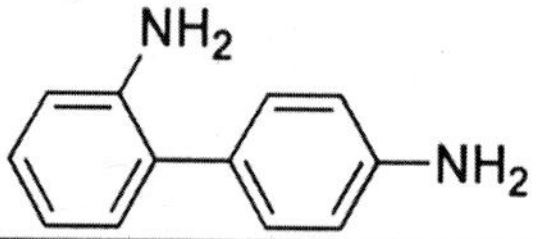	Not carcinogenic	Amine group in 2 position; missing 4-position
Biphenylamine 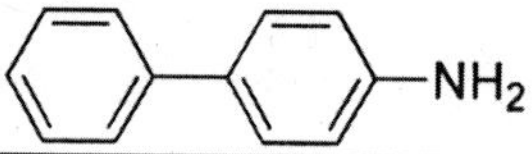	Carcinogenic	Not as potent as benzidine
Chrisoidine 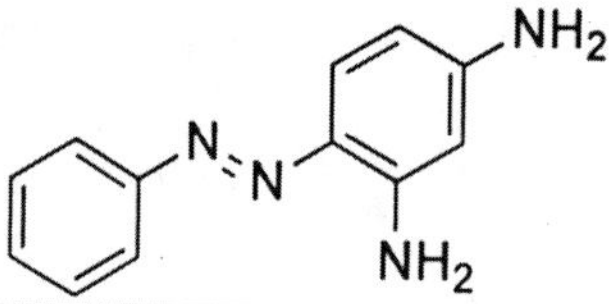	Not carcinogenic	Missing amine group in 4 position
4-Nitrobiphenyl 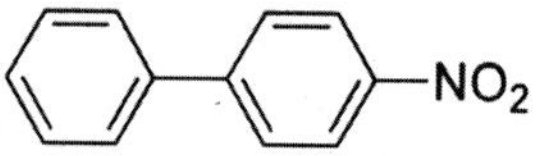	Carcinogenic	Nitro group in 4-position; amine group not required; other bioisosteres acceptable
2-Nitrophenol 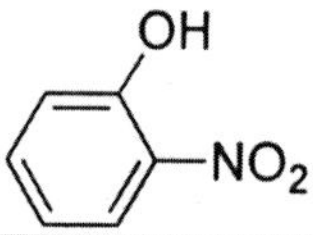	Not carcinogenic	4-position unfilled
Nitrofen 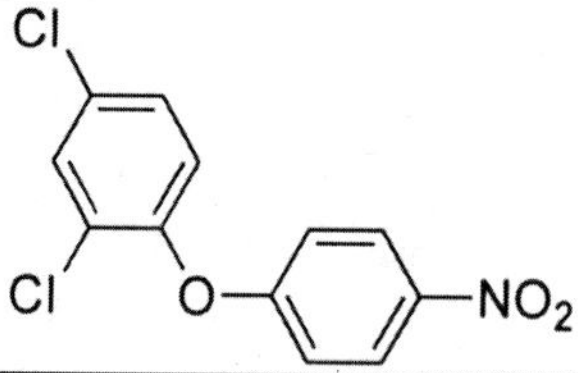	Predicted to be carcinogenic	Occupied 4-position

SOURCE: See reference 17.

The concept of "drug-likeness has been used to describe the molecular features that are common among the majority of therapeutic drugs that result in their effectiveness (*18*). The guidelines predict that poor absorption or permeation of an orally administered compound is more likely if the compound meets the following criteria:

- Molecular mass greater than 500 g/mol,
- logP greater than 5,
- More than 5 hydrogen bond donors, and
- More than 10 hydrogen bond acceptors.

These criteria are commonly referred to as the "rule of five" (*19*) and serve as an initial screen of new drugs. Structures of selected compounds can be presented to the students who must decide if the compound will likely be absorbed through the oral route using the guidance of the rule of five. For those compounds that are identified as likely being well absorbed according to rule of five guidance, students would suggest molecular modifications that would likely result in decreased availability leading to minimized hazard.

Lipophilicity is a measure of a chemical's affinity for the lipid bilayer of biological membranes. The logarithm of the partition coefficient between water and 1-octanol (log Kow) is used as an indicator of a chemical's lipophilicity. The parabolic relationship between log P and effect can be used as evidence that there are limits to absorption for "super-lipophilic compounds", and why these limits exist (*20*). Chemicals that have log P values greater than 6 tend to dissolve in the non-polar interior of a membrane inhibiting transport.

An instructor can ask the class as a discussion question whether the relationship between log P and toxicity is infinitely linear, in other words, is a compound with a log P value of 9 likely to be absorbed to a greater degree than a compound with a log P value of 5.

Distribution

Chemicals must be distributed to their site of action to manifest a biological effect. The distribution process is controlled by passive diffusion, active transport and protein binding in blood and tissues. In general smaller, hydrophilic molecules are distributed more quickly and to a greater extent to well-perfused compartments than are larger lipophilic molecules. Lipophilic molecules tend to become concentrated in slowly perfused compartments like bone and in fat stores.

The relative affinity of a compound for either the aqueous phase (plasma) or the lipophilic phase (fat and bone) will dictate where a compound is likely accumulated. An example that can be used to illustrate this is the time course of the accumulation of 2,3,7,8-tetrachlorodibenzo-*para*-dioxen (TCDD) in rats (*21*). Five minutes after intravenous injection of TCDD, the highly perfused lung contains 15% of the total concentration and the poorly perfused fat contains 1.0%. However, within twenty-four hours, the lung contains 0.3% and the fat 20%.

Metabolism

Metabolism includes all of the chemical transformations that occur in living systems. A detailed discussion of metabolism is beyond the scope of this chapter and the reader is directed to other comprehensive resources (*22*). The lesson to impart to all students interested in green chemical design is that fundamental chemical reactions are the foundation of all of biotransformation. Addition reactions, conjugation reactions, substitutions and eliminations occur to chemicals found naturally inside living systems (biochemistry) and to those that are found external to living systems (xenobiotics) as well as in laboratory round-bottom flasks. Table III lists some of the more common biotransformation pathways.

Phase 1 reactions refer to the introduction or unmasking of a functional group that makes a compound more water soluble or more able to react in a

Table III. Common Biotransformation Reaction Classes

Phase 1 Reactions (functionalization)	*Phase 2 Reactions (conjugation)*
Hydrolysis	Glucuronidation
• Epoxides	
• Aliphatic halides	Sulfonation
• carbonyl	
	Methylation
Reduction	
	Acetylation
• Azo and nitro reduction	
• Sulfoxide reduction	
• Quinone reduction	
• Aliphatic halides	
Oxidation	
• Aliphatic and aromatic hydroxylation	
• N-hydroxylation	

phase 2 reaction. These reactions can render a molecule less hazardous and more easily eliminated or it can activate a molecule resulting in greater hazard. It is therefore important for an instructor to emphasize that knowledge of the initial reaction and the potential products and associated hazards must be known to the extent feasible. Not all biotransformations are detoxicating.

Sulfanilamide is an effective antibiotic and has been used safely for many years to treat streptococcal and pneumococcal infections. However, sulfanilamide is generated from the azo-reduction of Prontesil, which is an example of a prodrug. A prodrug has no therapeutic effect on its own but generates active metabolites. The metabolism of prontesil, a prodrug, to its active metabolite, sulfanilamide, can be used as an example of azo reduction (Figure 6).

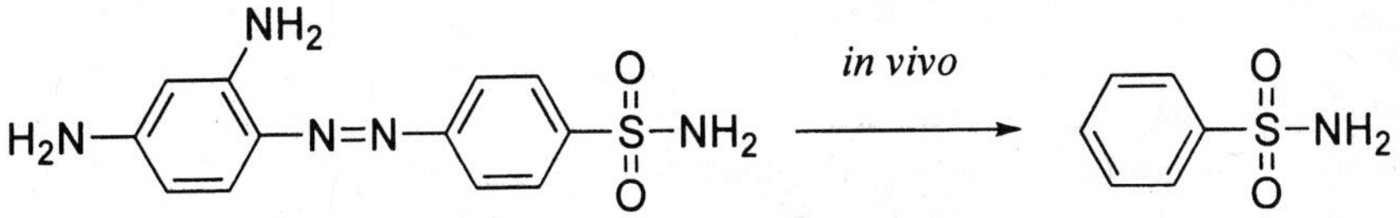

Figure 6. Azo-reduction of Prontesil to sulfanilamide.

Phase 2 reactions involve the conjugation of small molecules covalently to active compounds resulting in highly polar products. This increase in hydrophilicity enhances excretion through the kidneys. Examples of these reactions include glucuronidation, sulfonation, acetylation, methylation and conjugation with glutathione. The utility of understanding the metabolic fate of a chemical provides a synthetic chemist with valuable information from which to design a safer chemical. Predicting likely Phase 2 metabolic products based on an inspection of likely reaction pathways enables a molecular designer to potentially influence excretion and availability of new or existing chemicals.

Excretion

Toxicants are eliminated from the body by several routes including the kidney, lung and digestive system. The rate of elimination of chemicals that are primarily excreted through the lungs depends on volatility. Highly water-soluble chemicals are excreted through the kidneys and lipid soluble chemicals are generally eliminated through feces. As part of a discussion of the gas laws, the instructor can present the role of volatility and partial pressures on lung excretion.

Pharmacodynamics

Pharmacodynamics can best be described as what the chemical does to the organism and describes the effects on an organism after a chemical has reached its site of action, specifically the magnitude of the response (efficacy) or potency (the amount of the chemical that is needed to reach some maximal response). Even though the original focus of pharmacodynamics was to provide information on drugs, the concepts are universal and are therefore applicable to all chemicals. Pharmaco/toxicodynamics describe the effects of compounds at their site of action and the resulting biological effects. Ideally the description of events leading to the observable and measurable endpoint is at the molecular level, which is most useful to molecular designers.

Binding to specific receptors or non-receptor mediated effects (*i.e.*, narcosis) involve all of the interactions freshman chemists are familiar with including ionic, hydrogen and covalent bonding as well as van der Waals and hydrophobic interactions. Pre-medical and biochemistry students are a particularly well-suited, receptive target audience for this topic because of the direct application of this area to medicine.

Given adequate information about both the molecular structures and the biological actions of a group of congeners, it should be possible to identify the critical molecular features necessary for maximal activity. A well-modeled description of the relationship between structure and hazard allows for the informed design of novel compounds that possess reduced hazard.

Availability

In cases where virtually no understanding of the molecular basis for a particular chemical's hazard exists, there are still options for the molecular designer to reduce hazard. A chemical must interact with its target in order manifest hazard. This principle is true whether we are discussing interactions with a biological receptor, the ozone layer, or an ecosystem. Since this is the case, the principle of reducing the availability of the substance to manifest hazard is one that students must understand as a tool available to them. Whether it is designing a substance to reduce its ability to cross certain membranes in the body, or reducing chemical stability and volatility so a chemical cannot reach stratospheric ozone or attain its lower flammability limit concentration, molecular manipulation of availability is an essential tool for green chemical designers.

Conclusions

Green chemical design can be incorporated into existing curricula at several points in a high school, undergraduate and graduate chemistry program. Among chemists, physical hazards are the most familiar because of their acute nature and often dramatic and destructive results. All first year chemistry students appreciate the notion that chemicals can explode or burst into flames. The underlying reasons for a chemical having the ability or potential to explode or be flammable are more mysterious, which presents an opportunity to enlighten students on the structure-physical hazard relationship.

Introducing the fundamental elements of green chemical design, whether within the existing curriculum or through a separate course, is essential in training students to design environmentally benign products and processes. While certain concepts of toxicology (*e.g.*, dose-response) enhance a student's understanding of approaches to these design principles, it is not necessary to transform all chemists into practicing toxicologists in order to gain an appreciation for the impact that changes in structure can have on intrinsic hazard. Through the use of the fundamentals outlined above, future chemists can effectively achieve the goals of green chemistry and meet their responsibilities for introducing new substances into the world with minimal hazard to public health and the environment whenever possible. Not all chemists will become molecular designers, but all chemists should be given the tools to understand the intimate and fundamental relationship between molecular structure and hazard.

References

1. Collins, T. J. *J. Chem. Educ.* **1995**, *72*, 965.
2. Cann, M. C. *J. Chem. Educ.* **1999**, *76*, 1639.
3. Hjeresen, D. L.; Schutt, D. L.; Boese, J. M. *J. Chem. Educ.* **2000**, *77*, 1543.
4. Reed, S. M.; Hutchison, J. E. *J. Chem. Educ.* **2000**, *77*, 1627.
5. Beyond Benign Foundation. Green Chemistry: The Green Curriculum. http://www.beyondbenign.org/outreacheducation/kthru12.html (accessed Jun 20, 2008).
6. The Greener Education Materials (GEMs) for Chemists Database. http://greenchem.uoregon.edu/gems.html (accessed Jun 20, 2008).
7. *Designing Safer Chemicals: Green Chemistry for Pollution Prevention*; Garrett, R. L., DeVito, S. C., Eds.; American Chemical Society: Washington, DC, 1996.
8. Anastas, N. D.; Warner, J. C. *Chem. Health Saf.* **2005**, *12*, 9.
9. Anastas, P. T.; Warner, J. C. *Green Chemistry: Theory and Practice*; Oxford University Press: Oxford, U.K., 1998.
10. *Cassarett and Doull's Toxicology: The Basic Science of Poisons*; Klaassen, C. D., Ed.; McGraw-Hill: New York, 2001.
11. Albert, A. *Selective Toxicity: The Physicochemical Basis of Therapy*; Chapman and Hall: London, 1981.

12. Dieckhaus, C. M.; Santos, W. L.; Sofia, R. D.; Macdonald, T. L. *Chem. Res. Toxicol.* **2001**, *14*, 958.
13. Mandell, G. L.; Petri, W. A. Antimicrobial Agents: Penicillins, Cephalosporins and Other β-Lactam Antibiotics. In *Goodman and Gilman's The Pharmacological Basis of Therapeutics*; Hardman, J. C., Limbird, L. E., Eds.; McGraw-Hill: New York, 1996.
14. DeVito, S. Designing Safer Nitriles. In *Designing Safer Chemicals*; Garrett, R., DeVito, S., Eds.; ACS Symposium Series 640; American Chemical Society: Washington, DC, 1996.
15. Tanii, H; Hashimoto, K. *Toxicol. Lett.* **1984**, *22*, 267.
16. Selassi C. D.; Mekapati, S. B.; Verma. R. P. *Curr. Top. Med. Chem.* **2002**, *2*, 1357.
17. Shaw, I. C.; Chadwick, J. *Principles of Environmental Toxicology*; Taylor and Francis: London, 1998.
18. Ajay. *Curr. Top. Med. Chem.* **2002**, *2*, 1273.
19. Lipinski, C. A.; Lombardo, F.; Dominy, B. W.; Feeney, P. J. *Adv. Drug Delivery Rev.* **1997**, *23*, 3-25.
20. Hansch, C.; Steward, A. R.; Anderson, S. M.; Bently, D. *J. Med Chem.* **1968**, *11*, 1.
21. Weber, L. D. W.; Ernst, S. W.; Stahl, B. U.; Rotzman, K. *Fund. Appl. Toxicol.* **1993**, *21*, 523.
22. Parkinson, A. Biotransformation of Xenobiotics. In *Cassarett and Doull's Toxicology: The Basic Science of Poisons*; Klaassen, C. D., Ed.; McGraw-Hill: New York, 2001.

Chapter 9

Integrating Green Engineering into Engineering Curricula

Julie Beth Zimmerman[1] and Paul T. Anastas[2]

[1]Chemical Engineering–Environmental Engineering Program, School of Forestry and Environmental Studies, Center for Green Chemistry and Green Engineering, Yale University, New Haven, CT 06520
[2]Center for Green Chemistry and Green Engineering, Chemistry, School of Forestry and Environmental Studies, Chemical Engineering Department, Yale University, New Haven, CT 06520

The introduction of Green Engineering is taking place in colleges and universities around the U.S. and the World. Currently there are books that have been developed, courses and lecture materials being generated and a wide range of approaches to infusing Green Engineering Principles into the curriculum. This chapter will review the various approaches that are taking place and discuss specific techniques to introduce Green Engineering to students at both the undergraduate and graduate levels.

Introduction

Significant global challenges such as population growth, global warming, resource and water scarcity, ecosystem degradation, and environmental releases are gaining increasing attention and demanding increasing awareness. The importance, and growing urgency, of these problems are reflected in the fact that all global systems are now in decline from the oceans to the tropical rainforests to the mountain snowpacks (*1,2*). While these challenges are complex and far-reaching, there is a clear and critical role for the science and engineering communities to play in designing the next generation of molecules, products, processes, and systems to reduce or eliminate environmental impact while maximizing social benefit and performance.

Collectively, these challenges are part of a greater paradigm – sustainability – that has been defined as "meeting the needs of the current generation without inhibiting the ability of future generations to meet their own needs" (*3*). Among the critical components to promote a systematic shift towards addressing these challenges is increased sustainability and global awareness, such as the Principles of Green Engineering (Table I) (*4*), in engineering education. This shift towards sustainability education has been called for by the National Academy of Engineering in their *Engineer 2020* report (*5*) and the engineering accreditation body, ABET, in their undergraduate education assessment criteria (*6*).

While there have been some signs of progress of shifting engineering education towards the integration of sustainability principles – mutually advancing economic, environmental, and social goals – and a global awareness (*7*) at the individual department or university-level is being encouraged by the federal government through grants and recognition (*8*), fairly recent studies indicate that engineering curricula resist change (*9*) and continue to emphasize economic cost/benefit analyses (*10*). An additional challenge in transforming engineering education is the declining proportion of students choosing engineering as a profession. From 1994 to 2004, enrollment in U.S. higher education institutions has increased from 12.4M to 15.0M (*11*), while the number of degrees conferred in engineering fields have remained relatively flat (63.0K in 1994 and 64.7K in 2004) (*11*) with little diversity overall (*e.g.*, 22% female in 2003) (*12*).

Although the declining enrollments in engineering, minimal gender diversity, and a limited sustainability content in engineering curricula may appear to be disparate events, evidence suggests that these are in fact linked where sustainability curricula may actually increase the recruitment and retention of women and underrepresented groups in engineering as demonstrated through groups such as Engineers Without Borders (*13*); according to a survey of 3,332 teens worldwide, 14 to 18 year olds are committed to equality, prefer companies that act responsibly and care deeply about a clean environment (*14*); according to *The Women's Experiences in College Engineering Project* women leave engineering programs, not because of poor academic performance, but because curricula seem to have no connection to their goals of helping to improve society (*15*); International Senior Design courses offered at Michigan Technological University, which focuses on the engineer as a public-service agent, is 53% female (out of 118 students), which far exceeds the national averages for the enrollment of women in engineering. In short, these data suggest that young women and men are not choosing careers in engineering.
Technological University, which focuses on the engineer as a public-service agent, is 53% female (out of 118 students), which far exceeds the national

Table I. The 12 Principles of Green Engineering

Principle 1	Designers need to strive to ensure that all material and energy inputs and outputs are as inherently nonhazardous as possible.
Principle 2	It is better to prevent waste than to treat or clean up waste after it is formed.
Principle 3	Separation and purification operations should be designed to minimize energy consumption and materials use.
Principle 4	Products, processes, and systems should be designed to maximize mass, energy, space, and time efficiency.
Principle 5	Products, processes, and systems should be "output pulled" rather than "input pushed" through the use of energy and materials.
Principle 6	Embedded entropy and complexity must be viewed as an investment when making design choices on recycle, reuse, or beneficial disposition.
Principle 7	Targeted durability, not immortality, should be a design goal.
Principle 8	Design for unnecessary capacity or capability (*e.g.*, "one size fits all") solutions should be considered a design flaw.
Principle 9	Material diversity in multicomponent products should be minimized to promote disassembly and value retention.
Principle 10	Design of products, processes, and systems must include integration and interconnectivity with available energy and materials flows.
Principle 11	Products, processes, and systems should be designed for performance in a commercial "afterlife".
Principle 12	Material and energy input should be renewable rather than depleting.

SOURCE: Reference 4

averages for the enrollment of women in engineering. In short, these data suggest that young women and men are not choosing careers in engineering partially because this profession is perceived to lack a connection to helping improve the world around them. Although serving humanity is at the heart of the engineering profession and stated so in our ethics creed (*16*), the engineering education system and infrastructure (texts, learning aids, faculty development) have largely lost this core connection. Recognition of the challenges has led to innovations in engineering curricula around Green Engineering and Sustainability. While these advances are impressive and significant, there are still many opportunities to further these efforts and solidify Green Engineering as a standard and valued component of all engineering education.

Incorporation Of Green Engineering Into Engineering Curricula

The extent of integration of Green Engineering into engineering curricula at institutions of higher education may be identified by several key activities and indicators including but not limited to: (1) curricular innovations such as new core courses or electives or amending existing courses to include sustainability themes; (2) centers and institutes on campus related to sustainability; (3) conferences related to sustainability developed and hosted by faculty, departments, and schools; (4) institutional support and funding for research relating fundamentals, design and impacts on society, the economy, and the environment; (5) opportunities to pursue concentrations in departments focused on other aspects of sustainability including policy, economics, social sciences, business, and so on; (6) designated faculty with a single or joint appointment whose title, teaching, and research focus on sustainability; and (7) individual guest lectures, department or college-wide seminars or seminar series focused on sustainability (*17*). The following section will provide examples of current efforts in these areas.

Examples of Current Efforts

Development of curricula and educational materials

Through the development of curricula and educational materials, the ability to influence both teaching and learning is multiplied with benefits far beyond a single class or university. This can be achieved by the infusion of Green Engineering and sustainability into existing textbooks, developing new textbooks, and reworking laboratory and field experiences. Existing engineering textbooks have been updated to include discussions of sustainability, its relationship to engineering design, and the crucial role for the engineering profession in meeting the global challenges (*i.e.*, Mihelcic and Zimmerman, *Environmental Engineering: Fundamentals, Sustainability and Design*, John Wiley & Sons, 2008). Entirely new textbooks focused on Green Engineering have been developed and are under development to fill different niches in undergraduate and graduate courses as well as disciplinary needs. Due to the unique role of laboratory and field-based learning in engineering, this is an invaluable place to affect change by orienting these necessary and important components towards sustainability. A growing trend in this area is to introduce service-learning into field experiences in developing communities, both domestically and abroad. Of vital importance to these efforts is the ability for interested faculty to access these materials and integrate them readily into their courses without significant investments of time and energy. This suggests the need for curricular clearinghouses of engineering education materials focused on Green Engineering and sustainability that are accurate, modular, and optimized for the intended audience.

Centers and Institutes

Given the interdisciplinary nature of Green Engineering and sustainability, an emerging trend is to establish organizations that span disciplinary boundaries such as centers and institutes that can be within a single university or professional society or a collaborative effort between multiple schools and organizations. While many of these centers and institutes have been established with a research focus, nearly all of them recognize the importance of education and include a component focused on training the next generation. While centers and institutes established through professional societies lend credibility and visibility to Green Engineering and sustainability, the academic-based organizations play a unique role of hands-on work with students. Both of these mechanisms for operating and maintaining a center or institute are important and contribute to the successful integration of sustainability into engineering education.

Conferences

Green Engineering and sustainability in engineering education has been the subject of dedicated workshops as well as topics at several technical conferences. These events provide an opportunity for those focusing on Green Engineering and sustainability to convene and also provide a venue at large scientific meetings to raise the issue of sustainability and education. Workshops can bring together many different stakeholders including faculty, students, funding and accreditation agencies, and allow university administrators to focus on advancing synergistic opportunities and addressing common barriers. For example, on November 7 and 8, 2005, the US National Academy of Sciences hosted a workshop focused on this topic resulting in a report that highlighted the need, benefits, and potential challenges to integrating sustainability into science and engineering curricula (*17*). It is also important for these topics to be raised at technical meetings where those faculty conducting research in Green Engineering and sustainability understand the tremendous value in bringing these ideas into their classrooms. Both dedicated meetings and sessions at scientific gatherings are constructive and key to advancing the integration of Green Engineering and sustainability into engineering curricula allowing for a depth and breadth of exchange on this topic.

Institutional support / Research funding

Without institutional support for efforts to integrate sustainability into engineering curricula, it is very difficult for faculty to justify time and energy in this area, let alone succeed in this important task. This type of support can come in the form of academic, industrial, governmental, and non-governmental organizations coming together to support advanced training for students, establishing mechanisms for students to pursue opportunities in applying their engineering education to sustainability challenges, and funding opportunities focused on student research.

Groups, such as Engineers Without Borders and Engineers for a Sustainable World, were developed to encourage and support students who wish to enhance their education through service-learning and field-based research. The mission of these organizations is to match student education and research with a real world need in developing communities. These organizations have been extremely popular, realizing significant growth in the past few years with much of it coming from women and underrepresented groups (*13*).

Another program focused on sustainability education and research in engineering that has been steadily growing since its inception in 2004 is the US Environmental Protection Agency's P3 – People, Prosperity, and the Planet – Award program. This national student design competition supports student education and research on scientific, technical, and policy solutions to sustainability challenges in both the developed and developing world. EPA provides small grants to teams of interdisciplinary university students to pursue their projects and compete for larger grants to further develop their design, implement it in the field, or move it to the marketplace. One of the criteria used to make funding decisions is related to the use of the P3 Award program as an educational tool. Because one of the five key criteria used in evaluating proposals for funding is implementation as an educational tool, the P3 program has already exerted an exponential influence on the next generation of students in the schools' science and engineering departments. Since the design competition began, colleges and universities have integrated the P3 Award into their curricula in a variety of unique ways. Some have sent students to developing countries to educate and serve local communities. Other students have presented their projects at local conferences and professional seminars to educate members of the local community.

Similarly, if there is not research funding to study and assess the pedagogy surrounding sustainability training in engineering disciplines, then it is challenging to know which methods and approaches are most appropriate at achieving the desired learning outcomes. Both of these support mechanisms provide the foundation to successfully pursue education goals while developing effective strategies for dissemination of best practices and lessons learned in terms of integrating Green Engineering and sustainability into engineering education. The National Science Foundation has, through its Course, Curriculum, and Laboratory Improvement (CCLI) grants program, funded several projects focusing on assessing the various methodologies for integrating sustainability into engineering curriculum. This effort supports the development of best practices and dissemination of recommended approaches throughout the community. This will ultimately lead to more successful and valuable learning experiences for engineering students in terms of Green Engineering and sustainability.

Concentrations, Joint and Special Degrees

Given the complex and broad nature of the opportunities and challenges addressed by Green Engineering, it is becoming increasingly evident that it is necessary for students to have enhanced training and education in these areas and closely related disciplines (*18*). This has manifested in different approaches at different universities with several schools establishing programs for students to concentrate in green or sustainable topics, developing joint degrees, and even offering designated degrees in Green Engineering. While all of these approaches recognize that it is not effective or even practical to succeed in Green Engineering through a single discipline, there are potential costs and benefits associated with each strategy.

Students who pursue joint degrees are often required to take a heavier course load and may or may not find the experience isolating as it is can be difficult to establish a "home" department. For this reason, it is important that schools work to develop a critical mass of students pursuing the joint degree to ensure a sense of community for their students. At the same time, this path may or may not be valued by discipline-specific faculty who can be unsupportive of the student's objectives.

Another strategy for training the next generation in Green Engineering is to strategically integrate content relevant to sustainability throughout programs by redesigning existing courses and developing new ones. These programs train students to connect engineering design with the resulting impacts of these designs on human health and the environment. This approach raises visibility of Green Engineering and ensures that potential employers or graduate school admission committees are aware of the student's unique training particularly when this awareness is used to enhance a fundamental, rigorous engineering education. However, this approach when coupled with a change in program title to "Sustainability Engineering", "Green Engineering" or something similar also poses challenges in that 1) the community at large may not be familiar with this type of degree; 2) there can be a tremendous difference between one university's Green Engineering degree and another; and 3) the student may be perceived to be more a generalist than a fundamental and rigorously trained engineer.

Training Faculty / Designated Faculty

If the integration of Green Engineering into engineering curricula is going to be broadly successful, it is imperative that the faculty training the students have the opportunity to themselves learn about these fields as well as how to teach this information. Since Green Engineering was not prevalent when most of today's faculty went through their own training, several programs have been established to "train the trainers" in the Green Engineering and sustainability programs. These programs often also provide information and references to available education materials and modules that can be readily integrated into existing courses lowering the barrier to successful integration of Green Engineering. An added benefit of these programs is the creation of a network of faculty who are interested in integrating Green Engineering and addressing sustainability, facilitating information exchange and support.

Guest Lectures and Seminar Series

Another mechanism to bring Green Engineering to campuses is through guest lectures and seminar series. This provides an opportunity for an expert to present material that is most relevant to the department and may provide a perspective that is not currently represented on campus. Students and faculty alike often attend department-wide or topic-specific seminars allowing for opportunities of networking and collaboration around Green Engineering that may not have otherwise occurred. These seminars often have the added benefit of bringing together an interdisciplinary community from across the campus that may not interact under traditional circumstances. This has the potential to lead to collaborative research and education projects and enhance individual's efforts to bring Green Engineering and sustainability, more broadly, into the campus curriculum.

Looking Ahead

As demonstrated by all of these advances, there is a growing trend to significantly increase the integration of Green Engineering into engineering curricula. However, there are still many opportunities that could further advance these efforts such as overcoming institutional barriers including the tenure and funding systems that have a tendency to recognize traditional successes. Given the increasing enthusiasm and demand of students for this type of training, it is imperative that the current efforts are reinforced and new opportunities are sought to systematically and holistically integrate Green Engineering into engineering curricula.

References

1. Speth, J. G. *Red Sky at Morning: America and the Crisis of the Global Environment*; Yale University Press: New Haven, CT, 2005; p 320.
2. Brown, L. *Plan B 2.0: Rescuing a Planet Under Stress and a Civilization in Trouble*; W. W. Norton: New York, 2006; p 352.
3. The World Commission on Environment and Development. *Our Common Future*; Oxford University Press: New York, 1987.
4. Anastas, P. T.; Zimmerman, J. B. *Environ. Sci. Technol.* **2003**, *37*, 94A-101A.
5. National Academy of Engineering. *The Engineer of 2020: Visions of Engineering in the New Century*; The National Academies Press: Washington, DC, 2004.
6. Accreditation Board for Engineering and Technology (ABET). *Engineering Accreditation Commission, Criteria for Accrediting Engineering Programs—EC2000*; http://www.abet.org (accessed Jul 30, 2008).
7. Splitt, F. *Int. J. Eng. Educ.* **2004**, *20*, 1005-011.
8. Zimmerman, J. B. *Sustainability: Sci., Pract. Pol.* **2005**, *1*, 32-3.
9. Magee, C. L. *Needs and Possibilities for Engineering Education: One Industrial/Academic Perspective*; MIT Engineering Systems Division Working Paper Series, ESD-WP-2003-04; 2003, p 11; http://esd.mit.edu/wps/esd-wp-2003-04.pdf (accessed Jul 30, 2008).

10. Vanderburg, W. H. *J. Eng. Educ.* **1999**, *88*, 231-235.
11. National Science Foundation. WebCASPAR: Integrated Science and Engineering Resources Data System. http://webcaspar.nsf.gov (accessed Jul 30, 2008).
12. National Science Foundation, Division of Science Resource Statistics. Science and Engineering Degrees by Race/Ethnicity of Recipients: 1992-2001, 2004. http://www.nsf.gov/statistics/nsf04318/ (accessed Jul 30, 2008).
13. Zimmerman, J. B.; Vanegas, J. *Int. J. Eng. Educ.* **2007**, *23*, 242-253.
14. States, A. *Business Week* **2006**, Sept 18, 82.
15. Goodman, I. F.; Cunningham, C. M.; Lachapelle, C.; Thompson, M.; Bittinger, K.; Brennan, R. T.; Delci, M. Final Report of the Women's Experiences in College Engineering (WECE) Project, 2002. Goodman Research Group, Inc., Web Site. http://www.grginc.com/ExecutiveSummaries/GoodmanFinalPDF.pdf (accessed Jul 30, 2008).
16. National Society of Professional Engineers. NSPE Code of Ethics for Engineers. http://www.nspe.org/Ethics/CodeofEthics/index.html (accessed Jul 30, 2008).
17. National Academy of Sciences. *Exploring Opportunities in Green Chemistry and Engineering Education: A Workshop Summary to the Chemical Sciences Roundtable*; The National Academies Press: Washington, DC, 2007; http://www.nap.edu/catalog.php?record_id =11843 (accessed Jul 30, 2008).
18. Zimmerman, J. B.; Anastas, P. *Chem. Engineer* **2006**, *784*, 48-52.

Chapter 10

Green Laboratories: Facility-Independent Experimentation

Kenneth M. Doxsee

Department of Chemistry, University of Oregon, Eugene, OR 97403

By virtue of its focus on the reduction of intrinsic chemical risk rather than solely on minimization of exposure, Green Chemistry allows for laboratory investigations in settings that would be inappropriate for "conventional" chemical experimentation. The benefits of a Green curriculum are numerous, ranging from enhanced safety and cost savings to the facilitation of the (re)introduction of experimental chemistry, particularly at the K-12 and community college levels, where facility limitations have often curtailed laboratory investigation. A Green curriculum thereby promises enhancement of both the numbers and the diversity of students gaining knowledge of the practice of modern chemistry.

Overview

Many generations of students have euphemistically "had the opportunity" to carry out chemical experimentation in poorly equipped and inadequately ventilated spaces serving as teaching laboratories. However, modern laboratory safety procedures wisely limit such practice and its attendant risk of student exposure to hazardous chemicals. Recognizing reduction of exposure as an obvious means of reducing the risk of chemical investigations, most institutions attempt to provide well-ventilated laboratories, often having students perform all of their manipulations within the confines of efficient fume hoods. Many have also converted to microscale experimentation in order to further limit the likelihood of exposure to significant amounts of hazardous substances. Even in microscale experimentation, however, significant volumes of solvents are often evaporated in the fume hoods, and even small quantities of particularly hazardous substances can still present substantial risk upon exposure.

Risk Reduction via Minimization of Exposure

Unfortunately, avoidance of risk by reduction of exposure through use of "environmental controls" such as fume hoods presents several significant problems. Foremost among these is the fact that environmental controls can fail to protect students from intrinsically hazardous substances. How many of us have seen a student working literally *in* a fume hood, with head and shoulders inside the confines of the hood while adjusting an apparatus or adding solvent to the top of a tall chromatography column? (Indeed, how many of us have done this ourselves?) This all-too-common practice, of course, effectively converts the fume hood from a protective device into a fume concentrating device for the student. In addition, fume hoods suffer from the potential for outright failure due to obstruction of air flow *(e.g.*, by books, apparatus, or bottles, or by paper towels sucked into ventilation lines), to loss of electrical power, or to inefficient removal of particularly dense vapors (*e.g.*, bromine). Add to this picture the fact that fume hoods merely release vapors to the atmosphere atop the laboratory building, where building air intakes may be sited and upon which construction and maintenance workers may tread, and one may quickly conclude that reliance on fume hoods to reduce risk is significantly problematic. An EPA Survey (*1*) provides typical observations in these regards.

> "... a survey of the classrooms used by the high school science staff indicated that less than 40% of these rooms were equipped with chemical fume hoods. In addition, we found that none of the middle school science rooms or the secondary level art studios had been equipped with chemical fume hoods to provide additional mechanical exhaust. As a result, the construction of the individual classrooms limited what activities or materials could be safely used by the occupants without jeopardizing indoor air quality.
>
> During our inspection, we noted that most of the fume hoods were in a state of serious disrepair. We found units that were missing interior panels and covers and as a result leaked contaminants. We found hoods with malfunctioning or inoperative ventilation fans. We also observed hoods vented through duct work that was perforated. Finally, we also found

> that all the hoods serving the science area were constructed so that the hood exhaust was located adjacent to the hood intake. As a result materials evacuated by the hood could be re-entrained by the intake and brought back into the classroom."

All these concerns aside, fume hoods as environmental controls in the teaching laboratory represent a tremendously expensive approach to the avoidance of chemical risk. Typical university-level chemical teaching laboratory renovation or construction projects run into the many millions of dollars, with a substantial percentage of this sum devoted to the purchase and installation of fume hoods for each student laboratory worker. Once installed, the air handling costs rapidly begin to mount, not only through the electrical costs of the air handling equipment, but also with the fume hoods quickly and efficiently removing heated (in cold climates) or cooled (in warm climates) laboratory air and requiring the cost-intensive heating or cooling of "makeup" air. While the costs of the latter are not always recognized, they can be daunting. Illustrative is a recent study (*2-4*) of a University of California prototype energy-efficient fume hood design, which concluded that the new design, which reduced the airflow to 30% of that for a typical hood installation, "could save 360 gigawatt-hours of electricity in California, and 2,100 gWh in the United States." What cost savings would this reduction represent? "At $0.08 per kWh, the annual electrical savings per hood is about $1,000 (*4*)." On a more global level, the estimated energy savings if such reduced-flow hoods were adopted US-wide would be equivalent to approximately half the annual power output of the Bonneville Dam, on the Columbia River in the Pacific Northwest.

Green Chemistry: Reduction of Intrinsic Risk

What, then, does one do if one wishes to provide students with the opportunity to carry out chemical investigations in a laboratory setting? The answer, of course, is suggested by the appearance of this chapter in a volume dedicated to the concepts of Green Chemistry. Simply put, Green Chemistry approaches the issue of risk reduction not through minimization of exposure, but rather through reduction of intrinsic chemical hazards. By reducing the intrinsic hazards of the chemicals, solvents, and reagents being used, Green Chemistry reduces or eliminates the risk, even in the event of exposure, since exposure to an innocuous substance does not represent risk.

The advantages of the Green Chemistry approach to risk reduction are immediately obvious and realizable. Students enrolled in the organic chemistry laboratory program at the University of Oregon carry out bench-top experimentation in a laboratory that is strikingly simple in its design (Figure 1), comfortable and inviting due to its openness and low ambient ventilation noise levels, and safe not only given the intrinsic safety of the materials being used but also given the ease with which supervisory staff can observe all student workers. Additionally, the laboratory's open sight lines foster the free exchange of information among students. The architects and designers of this laboratory frequently marvel at the high level of student comfort and joy evident in this lab, particularly in contrast to the high student stress levels more frequently encountered in such a pre-professional class.

Figure 1. The Green laboratory at the University of Oregon.
(See page 4 of color insert.)

Laboratory Experience in Non-Traditional Settings

Although not an originally intended consequence of conversion to a Green curriculum, we rather quickly recognized that the intrinsic safety of Green Chemistry allows for experimentation in facilities that might not otherwise be viewed as viable laboratory settings. The extreme of this flexibility, perhaps, has been demonstrated by presentation of Green laboratory workshops in conventional meeting rooms in several resort hotels, including workshops held in conjunction with the 22nd Mexican Congress for Chemical Education (Ixtapa, Guerrero, Mexico, September, 2003) and with the 4th International Meeting on Chemistry Teaching at College and Pre-College Level (Mérida, Yucatán, Mexico, November, 2005). A typical scene from the Ixtapa workshop is depicted in Figure 2, in which student and faculty participants from throughout the Americas are seen working in a conventional meeting room, complete with blue cloth table coverings and floor-length draperies, clearly the antithesis of the typical organic chemistry teaching laboratory! Figure 3 shows a participant carrying out a solvent-free aldol reaction (*5*) while seated in a comfortable chair, again hardly a standard laboratory setting. Within ten minutes of conclusion of the Mérida workshop, the meeting room was in use as a lecture hall by a group of some thirty meeting attendees, not one of whom noted any air quality issues.

Contrast this situation with the not uncommon lingering chemical odors in a conventional teaching laboratory!

The concept of carrying out organic chemical experimentation in a hotel room is, of course, extreme. Or is it? The sad reality is that many, perhaps most, of the educational settings in which one would hope to provide the opportunity for hands-on chemical experimentation are seriously under-equipped to allow work with hazardous substances. In grade schools, middle schools, and high schools, in community colleges, and in colleges and universities, and particularly in developing nations, we are more likely to encounter poorly ventilated laboratories than we are state-of-the-art facilities. With many studies showing that interest in the sciences is highest in young children and progressively decreases with educational level (*6-9*), and other studies highlighting the value of hands-on experience and experimentation in retaining student interest in the sciences (*10*), the all-too-common inability for schools to offer laboratory experience given the lack of suitable, safe facilities demands attention.

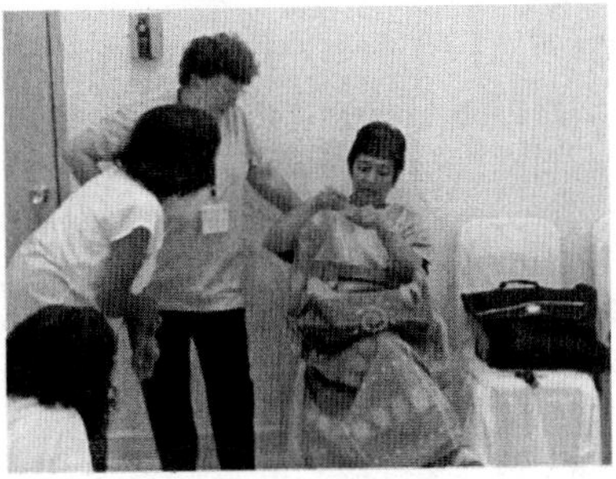

Figures 2 and 3. Organic experimentation in a hotel meeting room. *(See page 4 of color insert.)*

Green Chemistry, by permitting experimental chemistry to be performed in settings perhaps more closely approximating hotel meeting rooms than state-of-the-art laboratories, allows the reintroduction or revitalization of laboratory curricula throughout the full range of educational institutions, K-12 and beyond. The longer-term impacts of this change, in terms of the very face of chemistry, are staggering. Consider the implications at the community college level alone – a recent analysis (*11*) reports nearly 1,200 such institutions in the United States, enrolling over eleven million students, representing 45% of all US undergraduate students and 45% of first-time freshman students. Who are the students enrolled in community colleges?

- 32% are 30 years of age or older; 46% are 25 years of age or older.
- 59% are women.

- 85,000 are international students.
- 47% of black undergraduate students are enrolled in community colleges; 55% of Hispanic students, 57% of Native Americans, and 47% of Asian/Pacific Islanders.
- 50% of new nurses, and nearly 80% of emergency medical technicians are community colleges graduates.

Given fiscal realities, including historic under-funding and a need to maintain low tuition rates in order to ensure accessibility, many community colleges suffer from lack of availability of quality laboratory space. Can we afford to ignore these students, representing one of the greatest single sources of ethnic, gender, and age diversity, by not offering the opportunity to carry out state-of-the-art experimentation in the sciences? How many more students are impacted by the lack of laboratory facilities at the K-12 levels? How many more well-trained scientists would there be if we did not accept the inevitability of loss of interest in science with increasing grade level?

The Role of Green Chemistry in (General) Education

Is there a down side to Green experimentation? Critics occasionally object to the Green Chemistry curriculum, feeling that, by focusing on work with safe chemicals and procedures, students will not be properly trained to work with hazardous materials (*12*). We find this argument specious on a number of grounds. We do envision a future in which there may be no need for hazardous substances, and we do not see a problem with instilling this hope in the next generation of scientists, rather than pessimistically assuming a future in which chemical hazards are a given. As we work toward this vision, students trained in Green Chemistry learn about "peripheral" subjects such as toxicology and environmental chemistry. In the process, they develop the necessary tools to recognize and critically assess health and environmental hazards rather than following the common advice in traditional laboratories to "treat all chemicals as hazardous." In addition, "Green" is a relative term, and although some experiments are appreciably less hazardous than those they have replaced, they may still offer students the opportunity to work with things that are not completely safe. Finally, we must recognize our clientele – the vast majority of students enrolled in chemistry laboratory courses will, in fact, not become practicing chemists, but rather doctors, nurses, pharmacists, dentists, or any of myriad other professions. Most of them will not work with hazardous substances. They can be taught the scientific method and laboratory techniques with safe substances. Is it reasonable to "test" proper technique by seeing whether a student can successfully avoid being exposed to a known carcinogen? Importantly, the intrinsic interest of Green Chemistry also is attracting students who otherwise would be highly unlikely to consider chemistry courses. This new audience of students, who may go on to become lawyers, journalists, business owners, or politicians, will enter their chosen professions with a firm foundation in chemistry.

Summary

Green Chemistry reduces risk through reduction of the intrinsic hazards associated with a chemical or process, not solely through the reduction of exposure. Given its intrinsic safety, Green experimentation does not require special environmental control devices, with their concomitant installation and operation costs as well as the potential of accidental exposure as a result of either misuse or equipment failure. As a result, Green Chemistry allows for experimentation in a wide variety of settings, allowing the reintroduction of real experimentation as a key component in the scientific education of students at all educational levels. By allowing for enlivening of the science curriculum through laboratory investigation, Green Chemistry may play a key role in the education and training of a diverse future generation of scientists and citizens, fully reflective of the face of modern society.

References

1. Dresser, T. H. A Case Study of Environmental, Health & Safety Issues Involving the Burlington, MA Public School System. http://www.epa.gov/region7/education_resources/teachers/ehsstudy/index.htm (accessed Sep 7, 2008).
2. Mills, E.; Sartor, D. *Energy* **2005**, *30*, 1859-1864.
3. Bell, G.; Sartor, D.; Mills, E. *The Berkeley Hood: Development and Commercialization of an Innovative High-Performance Laboratory Fume Hood*; Lawrence Berkeley National Laboratory Report LBNL-48983 (rev.): Berkeley, CA, September 2003.
4. Lawrence Berkeley National Laboratory Applications Team. The Berkeley Hood. http://ateam.lbl.gov/hightech/fumehood/fhood.html (accessed Sep 7, 2008).
5. Doxsee, K. M.; Hutchison, J. E. *Green Organic Chemistry: Strategies, Tools, and Laboratory Experiments*; Thomson Brooks/Cole: Belmont, CA, 2004.
6. Simpson, R. D.; Oliver, J. S. *Sci. Educ.* **1990**, *74*, 1-18.
7. Greenfield, T. A. *J. Res. Sci. Teach.* **1996**, *33*, 901-933.
8. Jovanovic, J.; King, S. S. *Am. Educ. Res. J.* **1998**, *35*, 477-496.

9. Bazler, J. A.; Spokane, A. R.; Ballard, R.; Fugate, M. S. *J. Career Dev.* **1993**, *20*, 101-112.
10. Stake, J. E.; Mares, K. R. *J. Res. Sci. Teach.* **2001**, *38*, 1065-1088.
11. *National Profile of Community Colleges: Trends & Statistics*, 4th ed.; Phillippe, K. A., Sullivan, L. G., Eds.; Community College Press: Washington, DC, 2005.
12. Kramer, A. No more fuming at chemistry class. *Seattle Daily Journal of Commerce*, July 25, 2002; http://www.djc.com/news/en/11135656.html (accessed Sep 7, 2008).

Chapter 11

Student Motivated Endeavors Advancing Green Organic Literacy

Irvin J. Levy[1] and Ronald D. Kay[1,2]

[1]Department of Chemistry, Gordon College, Wenham, MA 01984
[2]Current address: Journal Production and Manufacturing Operations, American Chemical Society, Columbus, OH 43202

While great strides have occurred during the past decade in the areas of green chemistry and green chemistry education, informal surveys of undergraduate college science students indicate that a broad knowledge of this topic is still lacking. Nonetheless, we have observed in various venues that the green chemistry vision can be very motivational for typical undergraduate students attending organic chemistry courses. Our experience indicates that students can develop a powerful voice facilitating the paradigm shift toward green chemistry, simply by sharing the ideas of green chemistry with others. Here we report the activities of our Green Organic Literacy Forum (GOLum), in which organic chemistry students have developed and implemented projects designed to introduce green chemistry to a broad audience, and offer specific strategies to equip instructors to engage in similar outreach. We also discuss the impact of these activities on our students, our department, and our institution.

The Challenge for Green Chemistry

Recognizing the power and potential of chemistry to make a positive impact on the lives of people and society, in 2006 the American Chemical Society adopted a dynamic new vision statement (*1*):

Improving people's lives through the transforming power of chemistry.

This wonderful statement expresses a great spirit of optimism about the role chemistry can play in our society and provides a goal toward which the chemical enterprise should continue to strive in the decades to come.

In a similar way, proponents of green chemistry recognize the power and potential of this movement to make the chemical enterprise even more effective in achieving its goal of improving people's lives. In fact, the ACS vision statement suggests the following analogous vision statement for the green chemistry movement:

Improving chemistry through the transforming power of green principles and practices.

Indeed, the goal of the green chemistry movement involves transforming the chemical enterprise by changing the mindset of chemists, so that they will employ green principles in their practice of chemistry as a matter of course. When this transformation process is complete, we will no longer need to use the term "green" chemistry because chemistry itself will be green in an essential way. However, such transformations often take a long time—years or decades—and the green chemistry community faces the task of developing strategies to promote and hasten the process. As leading green chemistry educator, researcher, and advocate John Warner succinctly observed in a 2006 interview (*2*) with *The Boston Globe*, "The challenge for green chemistry is how do we change the thinking?"

Recent history suggests that it is possible for a significant cultural transformation to take place in a short period of time if there are significant catalytic events to promote that change. Consider, for instance, the explosive use of computing during the past three decades. In a very brief time, microcomputing evolved from a hobby (1978) to a business tool (1988) to a ubiquitous part of life in the United States (1998). In 2008, the quaint memory of an educational and research endeavor without microcomputing is increasingly dim. This transformation was sparked by the catalytic appearance, first to the business community, of electronic spreadsheet software such as Lotus 1-2-3 that empowered users to perform calculations common to the business domain. Implemention of the World Wide Web via the Internet was the final catalytic nudge that moved computing from the business and academic communities into the home. Clearly, advancements can occur through the slow and steady improvement of human knowledge and practice. However, for anyone who wishes to hasten those advancements, identifying effective catalysts is essential.

One Way to Catalyze Green Chemistry Education: "SMEAGOL"

In order to accomplish its task of transforming chemistry, the green chemistry movement must make progress on two fronts: shaping the mindset of the next generation of chemists and transforming the mindset of the current generation of chemists. In the same way that writing on a blank page requires less effort than changing a previously written page, training a new generation of chemists from the beginning to practice green chemistry is easier than changing the established mindset of the current generation, many of whom may be at best indifferent and at worst hostile toward calls to change "how we've always done it". Accordingly, as several of the chapters in this book report, significant progress on the education front has already occurred. While some headway has been made in exposing practicing chemists to green principles through initiatives such as the Presidential Green Chemistry Challenge Awards and green chemistry programs at selected chemical companies, these avenues may not effectively reach a broad range of other chemists, including educators at the high-school and undergraduate levels. Our work on green chemistry education and outreach at Gordon College provides one model for communicating the ideas of green chemistry to such communities: engaging students as advocates of green chemistry. We have called the concept underlying this approach *memetic catalysis*.

The term memetic catalysis derives its name from the word *meme*, which Dawkins (*3*) defined as a basic unit of cultural information. Passing memes from one person to another spreads information among the members of a community, and the rate of this information exchange grows as increasing numbers of community members receive the memes and pass them to others. If the memes represent transformational ideas, this process of memetic catalysis hastens the rate of transformation of the community. In the present case, the memes of interest are the principles and perspectives of green chemistry, and memetic catalysis takes place as proponents of green chemistry pass these memes to other members of the many communities of which they are part. Our experiences have shown us that our students can participate along with us in spreading the memes of green chemistry to communities of which they are part but which we as their teachers could never reach. Consequently, we have endeavored to empower students to join us in spreading the message of green chemistry by asking them to develop what we have called SMEAGOL: Student-Motivated Endeavors Advancing Green Organic Literacy.

This student-centered approach to promoting green chemistry has grown out of our own experience at Gordon College, a small faith-based, residential liberal arts college of 1600 students located in Wenham, Massachusetts, about 30 miles north of Boston. Our chemistry department is small and serves the cognate course needs of other departments (especially biology) while working with a small number of chemistry majors. In 2003, one of the biology majors taking the organic chemistry course, Laura Hamel, served as the catalyst for the transformation of our department into an enthusiastic proponent of green chemistry.

Throughout the 1990s many opportunities to learn about green chemistry began to appear. The following partial list gives a brief overview of the breadth of materials that were available to chemical educators by 2003:

- Books (*4*)
- Public Events (*e.g.,* Presidential Green Chemistry Challenge Awards) (*5*)
- Professional Society Publications (*6*)
- Scientific Journals (*7*)
- Education Journals (*8*)
- Specialty Conferences (*9*)
- Educational Resources (*10*)
- Educational Workshops (*11*)

Given this plethora of resources, it is somewhat surprising that such a small number of chemical educators were engaged in green chemistry education in the early 2000s. Like the vast majority of the chemical education community at that time, our department remained largely unaware of and indifferent to the principles and perspectives of the green chemistry movement.

Starting in 1986 and continuing each year into the early 2000s, every student in the second-semester organic chemistry course at Gordon College was required to produce a significant research paper on a self-selected topic of interest. Finally, after hundreds of these papers had been submitted over the course of almost two decades, Laura Hamel submitted a proposal in January 2003 to write her research paper on the topic of "Green Chemistry". Her choice of topic met with some resistance from her instructor (Levy), since he possessed little familiarity with the topic and had significant doubts about its suitability as a subject for the research paper. Nonetheless, despite his misgivings, he approved the topic but cautioned her that preliminary work was needed to determine whether the outcome she desired would be feasible.

As Hamel worked on her research paper, Levy attended the 225th National Meeting of the American Chemical Society, held in New Orleans in March 2003. At this meeting, a session titled "Sustaining the Earth through Green Chemistry" was scheduled for the Undergraduate Program. Since his student was preparing a paper on the same topic, Levy chose to attend this session, where he heard a presentation about green chemistry by Mary Kirchhoff, John Warner, Paul Anastas and Jim Hutchison—all leaders in the green chemistry community. Their inspirational lecture ended with an announcement about the Green Chemistry in Education Workshop (GCEW) to be held in August 2003 at the University of Oregon. Levy decided to attend the GCEW that summer, and he returned from there with a perspective transformed by the principles of green chemistry. While applying for the GCEW, Levy received the research paper "Green Chemistry: Environmentally Friendly Chemistry" from Hamel, and this work retains an honored place in the Gordon College Chemistry Department Archives as the introduction of green chemistry to our campus.

The critical point we wish to highlight in the above narrative is that a *student-to-faculty interaction* played the central role in sparking transformational change in the perspective of a faculty member and an entire department. As we have noted, many resources promoting an interest in green chemistry were available in early 2003, but these materials were largely ignored in our hands until a student's prompting caused us to give them serious consideration. This key idea arising from our department's actual experience has served as the basis of our efforts to empower students to catalyze the advancement of the green chemistry movement.

SMEAGOL at Gordon College: The GOLum Project

After attending the GCEW in August 2003, we became very interested in spreading information about green chemistry to audiences beyond our classrooms. Since direct interaction from our own students was the central motivational factor for us, our approach is based on empowering our current students to bring similar messages to audiences with which they already have a working relationship. These student-motivated endeavors advancing green organic literacy have taken a wide variety of forms at Gordon College, most of which have grown out of the organic chemistry course through the flagship program we have called GOLum: the Green Organic Literacy Forum.

GOLum, The Green Organic Literacy Forum

Starting in 2004, we replaced the traditional research paper in our organic chemistry course with a major outreach project (the "GOLum project"), which harnesses the enthusiasm and creativity of our students to spread information about green chemistry to audiences beyond our chemistry department. Informal entry surveys of students beginning organic chemistry indicate that they arrive with little prior knowledge of green chemistry. When asked to write a brief definition of "green chemistry," these students often reply:

- Chemistry related to plants or living things
- Chemistry of color
- New chemistry
- "I don't have any idea"

Indeed, our surveys of organic chemistry students from other institutions during 2003–2006 yielded only about 16% who could describe green chemistry as an environmentally friendly approach to chemistry. Interestingly, the data at Gordon College during the same period are very different, presumably because of the success of the GOLum outreach between our students and their peers as well as the support of other faculty in the department. In 2003, only 5% of Gordon College's incoming organic chemistry students could define green chemistry; however, after the first three years of the GOLum program, those numbers grew dramatically: 55% in 2004, 75% in 2005, and 70% in 2006.

Implementing GOLum

The ultimate goal of a GOLum project is to conceive, develop, and implement an effective presentation about green chemistry that will educate an external audience identified by the project team. This is not a simulated assignment; in order to receive full credit for their project, students must actually deliver their presentation to the target audience. Early in the second semester of the course, students begin their work on the project by forming a self-selected team of 4–5 participants. In order to become a team, all prospective team members must agree upon at least one day per week on which they can meet as a group, and each student must commit in writing to work together on GOLum for 1–2 hours each week on that day.

Once a GOLum team has formed, it must choose its target audience, which must be an actual audience to whom it can present its final work. Ideally this audience should have the potential to be affected in a significant way by the presentation. Examples of audiences identified by our students include the following:

- high-school students
- high-school educators

- college educators
- college students majoring in education
- parents of home-schooled children
- the broader campus
- the general public

In order to maximize the efficacy of the memetic catalysis, we encourage students to select an audience with whom they are already familiar. For example, a team making a presentation in a high school normally visits a school from which one of the team members graduated, returning to a former teacher's classroom. Though such a personal connection between the GOLum team and its audience does not always exist, it is our experience that such a relationship provides the greatest potential for genuine transformation of the audience's attitudes about green chemistry.

The next task for each GOLum team involves selecting a green chemistry topic to present to its target audience. This choice of topic is tightly connected to the identity of the target audience. For example, the material presented to a chemical educator would of necessity differ from that presented to a high-school chemistry student or to the general public. During the past four years, GOLum projects have covered a wide range of topics, such as the following:

- Introduction to green chemistry for high school via lecture and lab
- Development of greener organic chemistry laboratory procedures
- Campus outreach: Green Chemistry Through the Eyes of Faith
- Production of a green chemistry lecture series on campus
- Principles of green chemistry: The business perspective
- Development of a web-based green chemistry metrics calculator
- Green chemistry for education majors
- Production of a green chemistry video for high school students
- Production of biodiesel glycerin byproduct soap
- Research projects in green chemistry

While GOLum project topics vary widely, one shared characteristic required of every project is that it must have a potentially *sustainable impact* beyond the team's presentation. Too often, outreach presentations become like fireworks exhibitions that are brilliant to behold but quickly fade into memory. To prevent this from happening, each team must develop ways to ensure that its project continues to be useful after its formal presentation. For example, GOLum presentations to high-school chemistry classes have provided host chemistry teachers with resource materials related to the presentations so that the teachers can recreate the activities in subsequent years. Other projects have developed video or web resources that continue to be available to future GOLum teams.

Because of the complexity of the projects and the potential for group dynamics problems, several important milestones are required during this project. These milestones, and their approximate locations along the course calendar, include the following:

- *Team Organization*, Week 2. This milestone requires the formation of a team along with a work plan describing when weekly meetings will occur and what specific skills each team member brings to the project. Students are asked to brainstorm about an audience, topic and format and to present three or more possible projects.
- *Annotated Bibliography*, Week 5. This milestone requires the submission of a detailed list of resources that have been obtained for use in developing the GOLum project. This is an extremely important milestone as teams develop authority in their knowledge of the topic.
- *Forum Plan*, Week 6. The project team describes their specific target audience and their detailed plan for presenting their topic to that audience.
- *Peer-Critique of Forum Plan*, Week 8. This milestone requires an essentially complete project. However, instead of making the presentation for the target audience, the team makes its presentation in complete form to another team in the class. The evaluating team critiques the presentation using a guide sheet provided by the instructor.
- *Dress Rehearsal*, completed no later than Week 11. This milestone requires the team to present a dress rehearsal of the presentation for the instructor and invited guests in the week preceding the actual performance for the target audience.
- *GOLum Presentation*, completed no later than one week after the Dress Rehearsal. This milestone involves making the actual presentation to the target audience.
- *Final Submission of Project*, Week 12. In this final milestone of the GOLum project, the team submits to the instructor a binder containing all materials related to the project, including the audience description, the description of the presentation forum, the updated annotated bibliography, paper and/or electronic copies of all material presented to the audience, presentation assessments, analysis of outcomes, and suggestions for future work.

While our students have found their GOLum projects to be very time-consuming and demanding, the overall response concerning their participation in GOLum has been very positive. In particular, during the first three years of GOLum projects, 86% of students agreed that it was "useful to spread information about green chemistry to others", while only 3% disagreed with that statement. Similarly, over the first three years, 29% of students described the

GOLum project as an "excellent" assignment in the class, 43% "good", 19% "average" and only 9% described the experience as "poor" or "very poor".

The enrollment in second-semester organic chemistry at Gordon College is typically 35–40 students, which is divided into 7–10 GOLum teams. Our experience indicates that supervision of this many projects can be accomplished by one faculty member without difficulty. While the thought of conducting a GOLum-type program in a large lecture course is exciting because of the potential for green chemistry outreach represented by such a large number of students, direct supervision of the number of GOLum teams generated by such a course (40–50 teams for an enrollment of ~200 students) would be impossible for a single course instructor. Therefore, some creative administrative support arrangements would be necessary to implement this type of program in a large lecture course. For example, one or two GOLum teams could be given the task of overseeing other teams (*e.g.*, collecting and organizing submitted work, ensuring that deadlines are met, *etc.*) under the direct supervision of the instructor. Alternatively, course alumni could be recruited to supervise the teams [in a manner similar that used in the Peer-Led Team Learning model (*12*)], if stipend funding were available. Another possibility would involve employing a graduate teaching associate dedicated to managing the GOLum projects in cooperation with the instructor. Regardless of class size, however, the crucial task of the instructor in ensuring the overall success of the GOLum program is to energize and motivate the students by persuading them that their outreach efforts will make an invaluable contribution to the advancement of green chemistry.

Outcomes of SMEAGOL

For faculty who wish to engage in outreach for green chemistry, the GOLum project provides a way to amplify one's efforts, through the directed efforts of one's students. This not only allows the supervising faculty to accomplish far more than they could alone, but it also forges stronger collegial bonds between faculty and students. Anonymous comments from our students indicate that they have derived a great deal of personal satisfaction from the GOLum projects:

- "It is satisfying to know that we are helping spread the use of green chemistry, which will become increasingly important."
- "I have somehow contributed to the education of safe, green chemistry."
- "There will be a very positive result to our efforts, rather than just a research paper that gets filed away or thrown out."
- "I felt like I was an ambassador for something important."

Additionally, a number of GOLum projects have led directly to presentations by students at American Chemical Society meetings (*13*) as well as collaborative research projects between students and chemistry faculty. One of us (IJL) has received grant funding to develop a green chemistry education workshop in collaboration with a high-school chemistry teacher introduced to green chemistry by one of our GOLum teams. GOLum projects have also led to submissions to the Greener Educational Materials (GEMs) database (*14*). Our department has benefited from the goodwill engendered beyond our campus by the service our students have provided, and we have frequently been contacted with requests for assistance in green chemistry outreach activities in the Boston area. Finally, implementation of GOLum has assisted our department in recruiting and retaining majors and minors. During the first three years of the GOLum project, 43% of student participants anonymously responded in the affirmative that "I am considering adding a chemistry major or minor". While not all of those students actually became majors or minors in our department, a substantial number of them did so. Thus, even though this project was not designed as a recruitment or retention tool, we have seen the potential to leverage the experience of the GOLum project toward those ends.

Conclusion

While the GOLum project described in this chapter is a major component of our curriculum that requires a great investment of resources in time and energy, both for the students and the supervising faculty, we have found this experience to be invigorating because the projects have genuine outcomes. No longer do our students produce "term papers" that are quickly forgotten once a grade is assigned. Instead, the students work cooperatively to develop a sustainable project that can be useful in the greening of chemistry.

It should also be noted that faculty wishing to use smaller student outreach projects can also be extremely successful. For example, small assignments might include the preparation of posters to be placed around campus during National Chemistry Week each fall or the week of Earth Day each spring. To reach out to high-school educators, students might be asked to write letters to their former high-school chemistry teachers, explaining what they have learned about green chemistry and offering to provide further information about this topic. To reach out to the community, students could write articles describing the concepts of green chemistry that could be sent to family or friends or submitted to the college or local newspaper. In short, the range of projects is limited only by the creativity of those who assign them. The key theme is outreach to a community of which the information provider is already a member, thereby maximizing the potential for memetic catalysis and ultimately contributing toward improving chemistry through the transforming power of green principles and practices.

Acknowledgments

The authors thank the many students, colleagues, and administrators who have enthusiastically supported the green chemistry endeavors of the chemistry department at Gordon College. We especially thank Laura (Hamel) Ouillette for persisting after being met with resistance when she first brought green chemistry to our attention in 2003. The support of our friends in the Green Chemistry Education Network is phenomenal and has been a key to our success. We are grateful for the many folks who have become our colleagues and friends from among that group of apparently tireless advocates. I.J.L. particularly extends his thanks to his co-author, Ron Kay, who, as chair of the chemistry department at Gordon College, fully encouraged our transformation with green chemistry. Without his support throughout the past decade, this chapter could not have been written.

References and Notes

1. Raber, L. R. ACS Launches New Vision. *Chem. Eng. News* **2006**, *84* (13), 52-53.
2. Gavin, R. Making it easier to be green, 2006. Boston.com Business Web Page. http://www.boston.com/business/articles/2006/09/04/making_it_easier_to_be_green/ (accessed Nov 18, 2008).
3. Dawkins, R. *The Selfish Gene*; Oxford University Press: New York, 1976; Chapter 11.
4. For example: Anastas, P. T.; Warner, J. C. *Green Chemistry: Theory & Practice*; Oxford University Press: Oxford, U.K., 1998.
5. The Presidential Green Chemistry Challenge: Award Recipients 1996-2008. U.S. Environmental Protection Agency Green Chemistry Web Site. http://www.epa.gov/gcc/pubs/docs/award_recipients_1996_2008.pdf (accessed Nov 18, 2008), EPA document 744F08008, June 2008.
6. For example: cover stories in *Chemical & Engineering News*, July 16, 2001; July 1, 2002; June 30, 2003; July 12, 2004.
7. For example: *Green Chemistry*, published monthly by the Royal Society of Chemistry beginning in Jan 1999.
8. Collins, T. J. *J. Chem. Educ.* **1995**, *72*, 965-966.
9. For example: the Gordon Research Conference on Green Chemistry, held biannually beginning in 1996, and the Green Chemistry and Engineering Conference, held annually beginning in 1997.
10. For example: *Introduction to Green Chemistry*; Ryan, M. A., Tinnesand, M., Eds.; American Chemical Society: Washington, DC, 2002.

11. For example: (a) Summer School on Green Chemistry, Interuniversity National Consortium "Chemistry for the Environment" (INCA). http://www.incaweb.org/education/summer_school_on_green_chemistry/, (accessed Nov 18, 2008). (b) Green Chemistry in Education Workshop, University of Oregon. http://chemistry.gsu.edu/CWCS/green.php (accessed Nov 18, 2008).
12. The Peer-Led Team Learning Workshop Project Home Page. http://www.sci.ccny.cuny.edu/~chemwksp/index.html (accessed Nov 18, 2008).
13. (a) Kwon, S. Y.; Levy, I. J.; Levy, M. R.; Sargent, D. V.; Weaver, M. A. Measuring Ecotoxicity: Green Chemistry Experiment for the Undergraduate Laboratory Curriculum. *Abstr. Pap.—Am. Chem. Soc.* **2008**, *235*, CHED-352. (b) Langlais, A. L.; Levy, I. J.; Wagers, A. L. Green Chemistry Curriculum Development to Empower Outreach to Middle- and High-School Students and Teachers. *Abstr. Pap.—Am. Chem. Soc.* **2007**, *234*, CHED-121. (c) Luhrs, A. R.; Stoeckle, J. J.; Hendrickson, D.; Schiffer, A.; Levy, I. J. Green Chemistry Metrics Calculator (GCMC): A JavaScript Application for Green Chemistry Education. *Abstr. Pap.—Am. Chem. Soc.* **2006**, *231*, IEC-148. (d) Austin, M. R.; Laporte, K. L.; Levy, I. J. Synthesis and Photochromism of 6-Nitrospiropyran: A Green Chemistry Preparation for Organic Chemistry Laboratory. *Abstr. Pap.—Am. Chem. Soc.* **2006**, *231*, CHED-283. (e) Wetter, E.; Levy, I. J. Multicomponent Zeolite-Catalyzed Synthesis of β-Acetamido Ketones: A Green Chemistry Laboratory Experience. *Abstr. Pap.—Am. Chem. Soc.* **2006**, *231*, CHED-272. (f) Hamel, L. J.; Levy, I. J. Analyzing Green Chemistry: A Description of Metrics with Applications in Academia and Industry. *Abstr. Pap.—Am. Chem. Soc.* **2005**, *229*, CHED-1364.
14. The Greener Education Materials (GEMs) for Chemists Database. http://greenchem.uoregon.edu/gems.html (accessed Nov 18, 2008).

Chapter 12

K-12 Outreach and Science Literacy through Green Chemistry

Amy S. Cannon[1] and John C. Warner[2]

[1]Beyond Benign Foundation, 66 Cummings Park, Woburn, MA 01801
[2]President and Chief Technology Officer, Warner Babcock Institute for Green Chemistry, 66 Cummings Park, Woburn, MA 01801

Green Chemistry is a call to arms for the next generation of students to study the physical sciences. The philosophy of green chemistry puts a subject, which is generally considered abstract and difficult, into a familiar context relevant to the daily lives of students. The practice of green chemistry ensures a sustainable future with safer alternatives to chemicals products and processes. Within the United States there is a general decline in the percentage of students studying in the physical sciences. The message of green chemistry resonates with students and can inspire students to pursue the sciences. Green Chemistry materials and programs are needed at all educational levels in order to provide content for learning about the field. Beyond Benign, a non-profit dedicated to green chemistry education and outreach, is actively involved in K-12 outreach and curriculum development and training. This chapter describes three ways Beyond Benign is engaging teachers and students with green chemistry: through interactive classroom visits, curriculum development and teacher training. By providing materials and training at the K-12 level, students and teachers alike can be engaged in the subject of chemistry and learn about the science within a sustainable framework.

The general public

"I am a chemist." Speaking those words in certain circles can elicit all kinds of responses. There is a general misunderstanding and misconception about what it means to be a chemist. The portrayal of a chemist in the eyes of the public generally includes a person with wild hair, holding a bubbling flask and wearing a lab coat. Another image that comes to mind is that of explosions and fires. Why do people have this misconception? As chemists, we do not do an effective job of communicating to the general population. When we perform demonstrations to a group of school children to get them excited about science we do exactly what we would hope never happens in the "real world": explode things and set them on fire. Therefore, it is no wonder that when the general population thinks of chemists they believe that we do precisely what we do when we perform demonstrations.

There are many other reasons for the negative portrayal of scientists and chemists. Part of this is due to a general lack of science literacy in the public and therefore many images of scientists are distorted. Therefore, when a young person sits down and dreams about what they will do in their future careers, one of the last things they are thinking of is chemistry, mostly because they have misconceptions about what it means to be a chemist or they simply have no idea what a chemist does.

As chemists, we try to circumvent the negative image through advertising campaigns, which speak about the positive aspects of chemistry and how it affects our daily lives. For example, Dow recently launched the "Human Element" campaign; the American Chemistry Council has their "Essential2Living" campaign and DuPont has changed their slogan "Better living through chemistry" to "the miracles of science". These slogans and advertising campaigns help to improve the overall portrayal of science in the eyes of the general public. However, the negative image of chemistry must be addressed at the earliest levels, in K-12 education, in order to have the greatest impact on changing the portrayal within society.

Student enrollment in the sciences

According to the National Science Foundation (NSF), in 2004 just over 14,000 bachelor's degrees were awarded in the physical sciences in the United States. This number represents 1% of all bachelor's degrees awarded in the U.S., down from 3% in the 1960's (*1*). At the bachelor's level over the past 40 years, the absolute number of students graduating with degrees in the physical sciences has also dropped (Figure 1).

Taking a closer look at chemistry degrees, the NSF reports in its latest study (January 2007) a gradual decline in the number of chemistry undergraduate

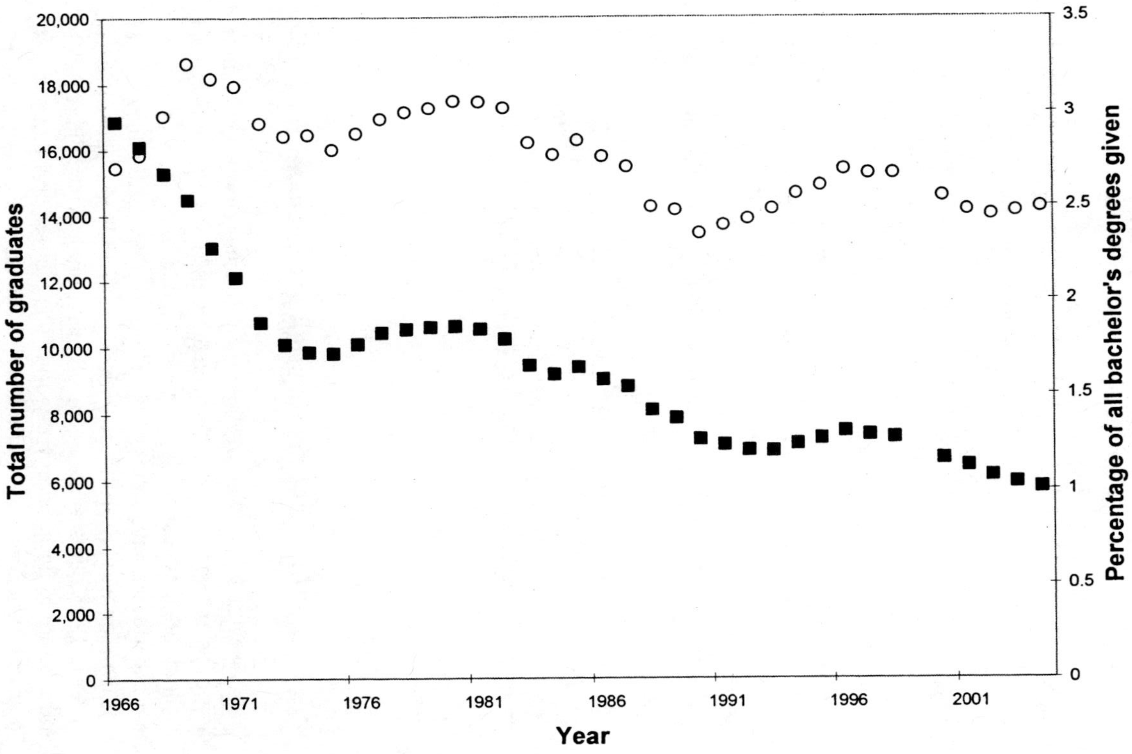

Figure 1. The degrees given in the physical sciences in the U.S. from 1966-2004, represented as: (○) the total number of graduates and (■) the percentage of all bachelor's degrees given, as reported by the National Science Foundation.

degrees awarded over the past 40 years (Figure 2). The NSF statistics differ slightly from the American Chemical Society (ACS) reports, which show the number of graduates in chemistry gradually increasing from 1971 to 2005 (*2*). The difference in the NSF and ACS studies can be attributed to the type of degrees that are reported in the studies; the NSF reports degrees in only the classic chemistry sub-disciplines, while the ACS reports degrees in biochemistry along with the chemistry degrees (*3*). The reported increase in chemistry degrees by the ACS might be attributed to the increased interest in biotechnology and the biological sciences over the past three decades.

Despite the slight variation in studies, the overall percentage of degrees given in chemistry is on a descent when compared to the total number of bachelor's degrees awarded in all disciplines (Figure 3) (*1-3*). This small percentage of degrees awarded in chemistry leads to only a small percentage of the population who have basic undergraduate training in the field of chemistry. In 2005, about one third of 25- to 29-year-olds had completed a bachelor's degree (*4*). And, of those bachelor's degrees, less than 1% of them were awarded in chemistry. Therefore, a very small percentage, less than one third of a percent, of the U.S. population has some training (bachelor's degree) in the chemical sciences.

The trend in science understanding in elementary and secondary education is also downward as the students advance through their education. The National Assessment of Educational Progress (NAEP) at the U.S. Department of Education reports a decline in basic level science understanding by grade 12 at which point close to 50% of the student population has a "below basic" understanding of science (Figure 4). In grade 12 only about 18% of the student population represents an "advanced" or "proficient" understanding of science. This number has decreased by 4% since 1996 (*5*).

Addressing the concerns through green chemistry

How can we address the decline of student enrollment, student understanding of science and the overall perception of science? The late Nobel Laureate Richard Smalley called for a new "Apollo program" in order to increase the number of students studying the physical sciences. Smalley's call to the younger generation was "Be a Scientist, Save the World" (*6*). This call seems to resonate with K-12 students as they generally have an understanding of global problems and a desire to do good in the world. The fact remains that chemistry and the physical sciences are intimately integrated into our society. Environmental damage and hazards do not have to go hand-and-hand with chemistry; there are many successful examples of green chemistry, which have illustrated this time and time again. These examples should be celebrated and used to demonstrate that chemistry affects our day-to-day lives in a positive

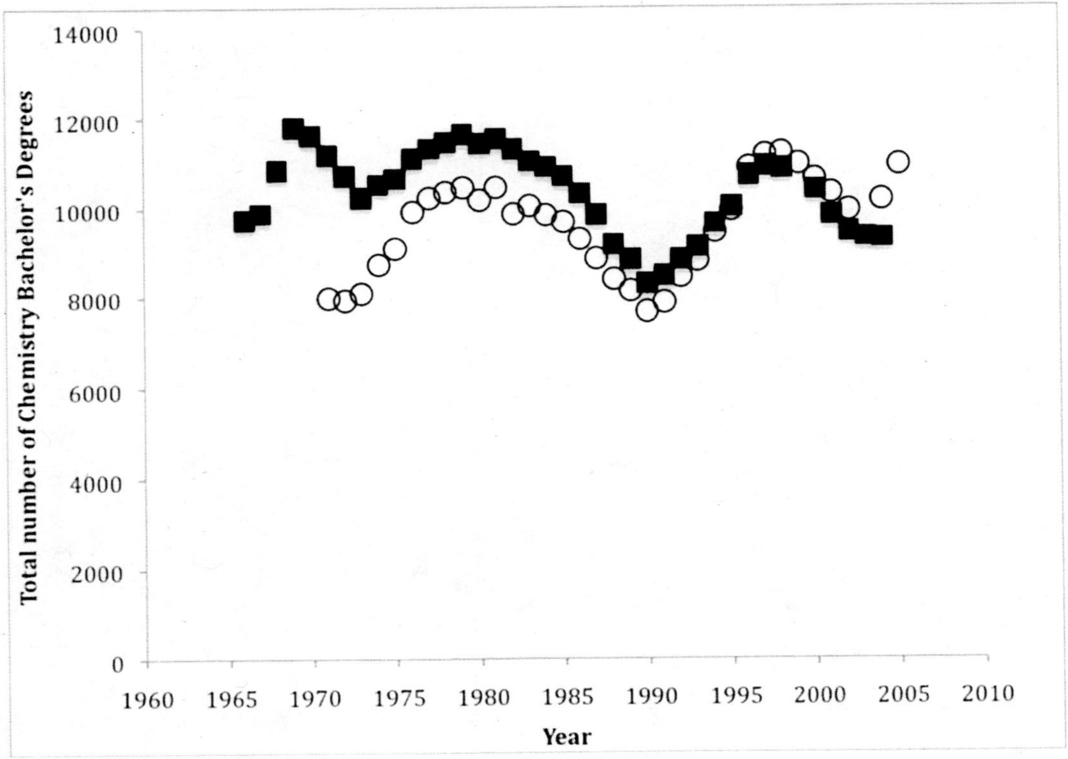

Figure 2. The degrees given in chemistry in the U.S. from two studies: (○) the American Chemical Society (■) the National Science Foundation.

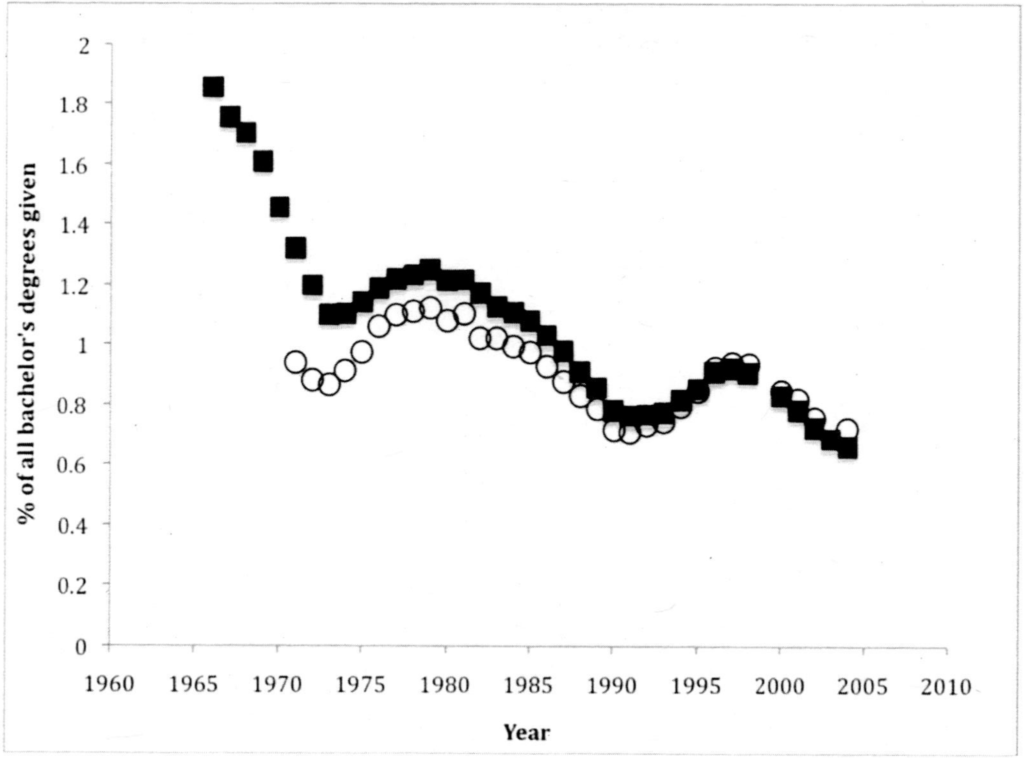

Figure 3. Chemistry degrees awarded in the U.S. as a percentage of all bachelor's degrees awarded according to: (○) the American Chemical Society (■) the National Science Foundation.

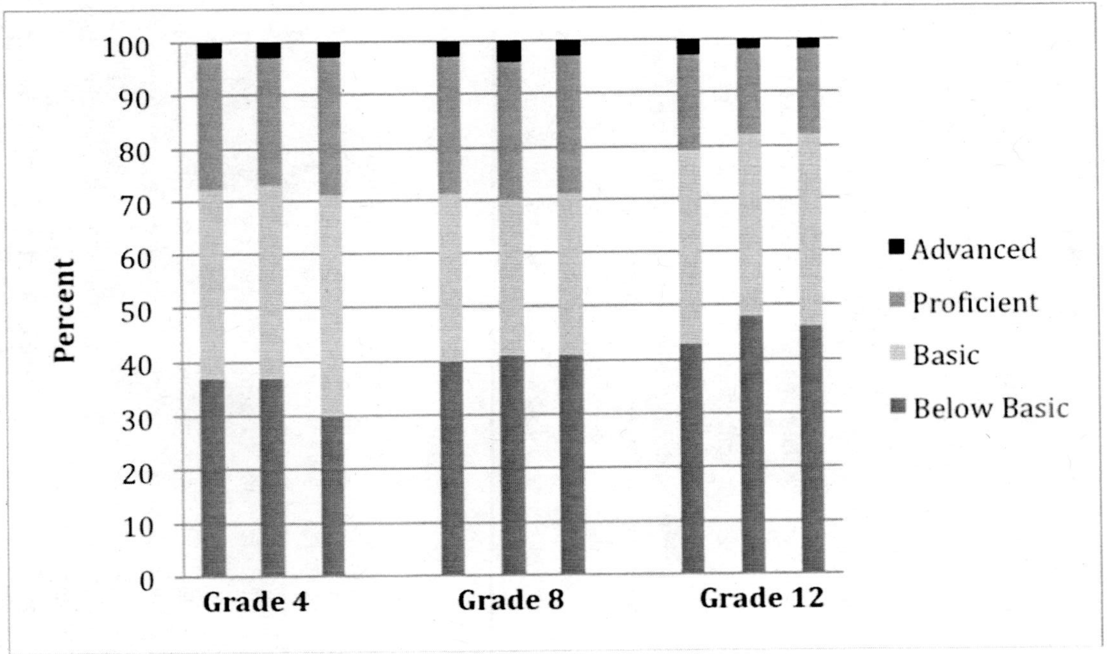

Figure 4. The NAEP science achievement-level performance at grades 4, 8 and 12 (5).

manner. Green chemistry can be a call to the next generation of students to help invent safer alternatives for our global society.

Green chemistry K-12 outreach

The community of Green Chemistry educators has an interest in engaging students at an early age to teach them about green chemistry. The goal is to inspire the next generation of scientists and also to teach students about chemistry and how it relates to their day-to-day lives. Not every child will be inspired to study the physical sciences, but all will grow up to be consumers and voters and can have an impact on green chemistry implementation.

Beyond Benign, a green chemistry educational organization, actively engages in K-12 outreach with the community. The overall organization is a non-profit taskforce focused on promoting science literacy in the interdependent arenas of community, industry and education in order to create a safer and more sustainable world. Driven by the principles of Green Chemistry, Beyond Benign creates tools, opportunities and partnerships to support the implementation of community involvement initiatives, workplace training and cooperation programs, and K-12 education resources. The Green Chemistry K-12 outreach initiative at Beyond Benign focuses on three main areas: 1) interactive K-12 visits, 2) K-12 teacher training and 3) K-12 curriculum development.

Interactive K-12 visits

The theme of the K-12 visits is "innovation through green chemistry." Technologies which have been demonstrated to be beneficial from a green chemistry perspective must have not only environmental benefit, but also the efficacy and economics to be implemented into the marketplace. Therefore, the academic and scientific rigor to achieve green chemistry's goals must be extremely high. By explaining green chemistry within the context of developing safer products that also have performance functionalities and economics, students begin to understand the impact of developing green chemistry technologies. K-12 students see that they can take part in the invention process and help create the next generation of products and materials that are economical, efficacious and environmentally friendly.

During the interactive visits, hands-on experiments are done with the K-12 students, guided by Beyond Benign staff and student volunteers. Inquiry based experiments generate discussions of innovation and green chemistry. Three hands-on experiments used regularly are solar cell construction, photoresist preparation using a DNA mimic, and photochromic spiropyrans. The three experiments are described in detail in the following pages.

Solar Cell Construction

In this experiment, students construct their own dye-sensitized solar cell using a titanium dioxide (TiO_2) semiconductor, blackberries as the sensitizer dye, graphite from a pencil as the catalyst for the counter electrode, Parafilm as the insulating seal and an iodine solution as the electrolyte (*7*). The experiment gives an opportunity to discuss many issues, including alternative energy, photosynthesis and redox chemistry. At the end of the experiment the students test how well they put their device together and determine how much energy they are able to harvest from the sun.

There is a great demand for alternative energy in today's society. Solar energy is a promising alternative to fossil fuels, but the construction of the devices uses large amounts of energy and potentially hazardous materials, therefore increasing the cost. A relatively new type of solar energy device, dye-sensitized solar cells (DSSCs), have potential to use less energy in the construction than that required in conventional solar cells. DSSCs are based on different chemistry than the traditional silicon-based technologies (*8*). The Warner research group has been investigating low-energy, low-toxicity methods for preparing DSSCs.

A cross section of a DSSC is shown in Figure 5. The solar cell consists of two electrodes, which have a thin transparent conductive coating of indium-tin-oxide (ITO) on the surface. One electrode has a TiO_2 semiconductor coating. The TiO_2 is dyed in order to sensitize the semiconductor to the visible spectrum. The counter electrode is coated with a catalyst (typically platinum or carbon) which functions to promote the generation of I^- in the electrolyte solution. The electrolyte solution reduces the oxidized sensitizer dye to its original state. The insulating seal keeps the liquid electrolyte intact.

The major component of DSSCs is the semiconductor, which is typically a metal oxide such as TiO_2. TiO_2 is non-toxic and inexpensive and has potential to be processed at low temperature. Traditionally, TiO_2 semiconductor films are made by preparing a sol-gel solution of the TiO_2, coating it on the glass substrate and sintering it at temperatures of 450-500 °C (*9–11*). The high temperature needed for these films limit the type of substrate that can be used in the construction; that is, the semiconductor cannot be coated on a plastic substrate, which requires lower processing temperatures. Through bio-inspiration, we have found that by adding a small amount of an additive during the sol-gel process, the films can be processed at much lower temperature (80 °C), providing high quality films with the same physical characteristics as the traditionally prepared materials (*12*). This process mimics that of the mineralization of shells and bones where proteins and small molecules chaperone the calcification process. The lower processing temperature translates to greater flexibility in coating substrates and lower cost. The generation of current from a DSSC has three steps (Figure 6):

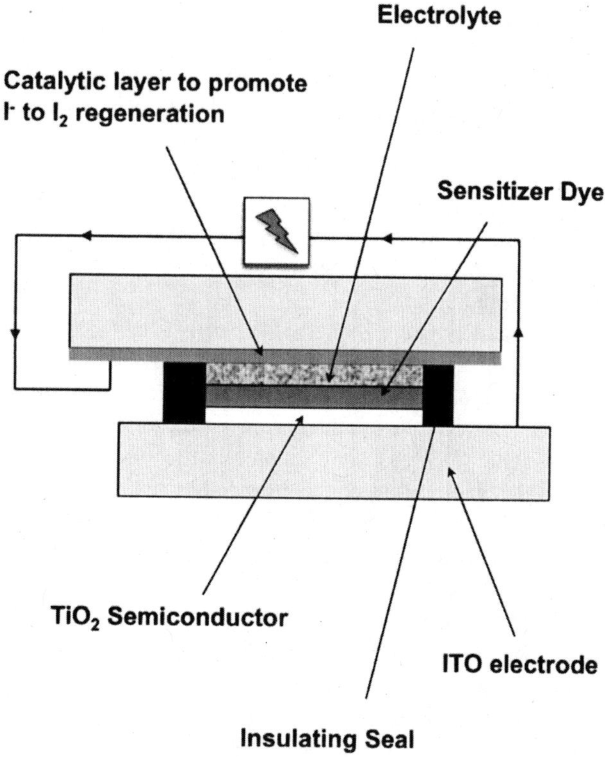

Figure 5. Cross-section of a DSSC.

1. Dye, adsorbed on a layer of semiconductor (TiO_2) interacts with the visible light provided by the sun (just like the green pigment does in a leaf), promoting an electron from a lower level orbital to an excited one.
2. The excited electron is injected by the dye into the semiconductor and travels to the outside circuit.
3. The electrons return to the cell to complete the circuit and bring the dye back to its ground state by using an electrolyte solution that helps carry electrons through the cell (typically I^-/I_3^- redox couple).

Photoresist preparation using a DNA mimic

In this experiment students prepare a photoresist thin film. A photoresist is a polymer that when irradiated with light will resist dissolution. The photoresists can be used for many applications from electronic devices to personal care products (*13*). They are used extensively in electronic materials. For example,

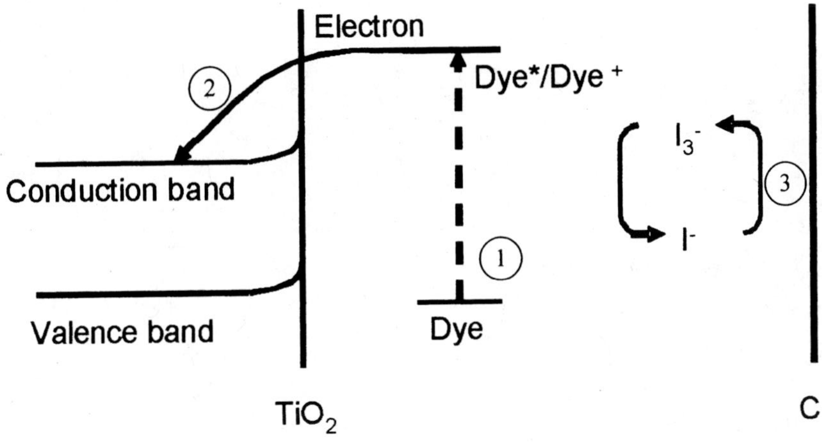

Figure 6. The mechanism of a DSSC.

within a computer keyboard is a pattern that maps the keyboard, which is made using a photoresist. Traditional photoresist materials are based on acrylate monomers, which are highly reactive molecules and tend to be highly toxic. There has been much effort spent working with alternative photoresist materials, which are non-toxic and water soluble (*14*). In this experiment, students coat their own photoresist, pattern it using light, and then dye their photoresist in order to see the pattern they have constructed (*15*).

A photoresist developed by the Warner research group mimics a process that occurs in DNA. Upon irradiation with ultraviolet light, thymine in DNA will undergo dimerization, coupling neighboring thymine moieties in the DNA chain (Figure 7) (*16, 17*). This reaction renders DNA nonviable, leading to mutations during replication that are potentially carcinogenic (*18*).

In humans the cross-linked thymine moieties create a kink in our DNA. An excision enzyme recognizes this kink and that portion of DNA is cut out and repaired. Skin cancer has been linked to the failure of this mechanism. Another repair mechanism is found in some microbes which have an enzyme called DNA photolyase. DNA photolyase functions by recognizing the thymine dimers and "un-zipping" them (*19*). We have found methods to use DNA photolyase as an enzyme for regenerating thymine polymer photoresists, allowing re-use of the photoresist systems (*20*).

Synthetic polymers that incorporate thymine take advantage of this dimerization reaction. Water-soluble polymers can be made that contain thymine in the chain. When the polymer is irradiated with UV light, it is transformed into a material that resists dissolution in water. In this experiment a water-soluble thymine based photoresist is prepared. The basic procedure is outlined in Figure 8. The water soluble photoresist is coated onto a substrate such as a plastic film.

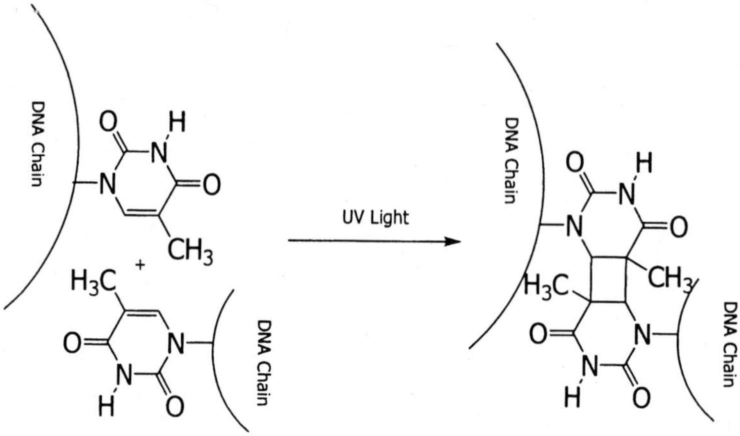

Figure 7. Photodimerization of thymine in DNA.

A blocking mask is placed over the coating and it is then exposed to UV light. The photo-dimerization reaction occurs in the regions where UV light hits, therefore rendering those regions water insoluble. The photoresist is washed with water to rinse off the unreacted portions. The photoresist is then patterned with a dye to make the film visible.

When compared to conventional photoresists, the thymine-based photoresists have many environmental benefits while conventional photoresist processes have many environmental and health concerns. Typical acrylate materials are organic solvent dependent monomers that undergo polymerization upon irradiation. These often toxic monomers are collected in the organic wash stage, requiring strict monitoring of waste and solvent evaporation. The process can be visualized by imagining many small individual molecules that undergo conversion into a huge networked material (Figure 9).

Thymine-based polymers are advantageous for several reasons. First, they are water-soluble, which avoids the need for organic solvents, an environmentally beneficial objective on its own. Second, a polymerization reaction is not necessary. These water-soluble non-toxic polymers are already polymerized. The photoreaction initiates a cross-linking mechanism by which neighboring strands are "tied" together (Figure 10). The formation of networks in this way makes them insoluble.

This experiment allows for the discussion of many topics including plastics and polymers, UV light, skin cancer, and DNA. K-12 students are able to see the preparation of a photoresist material and see how it applies to their day-to-day lives.

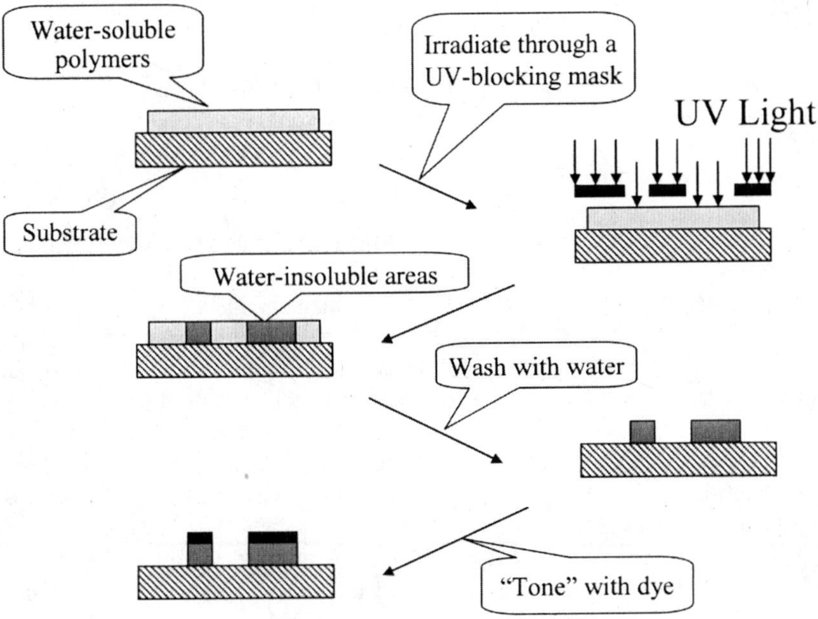

Figure 8. Preparation of a photoresist.

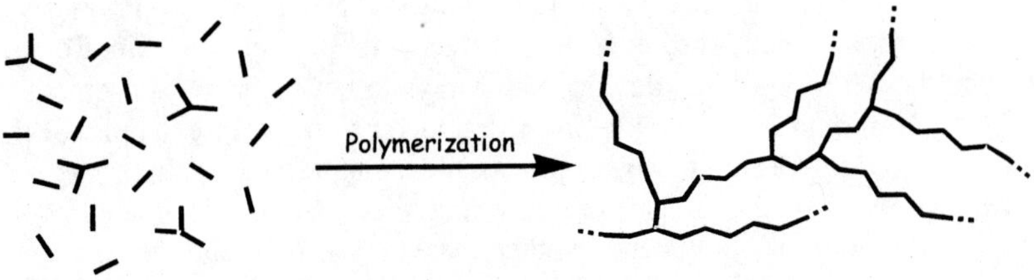

Figure 9. Traditional photoresist polymerization process.

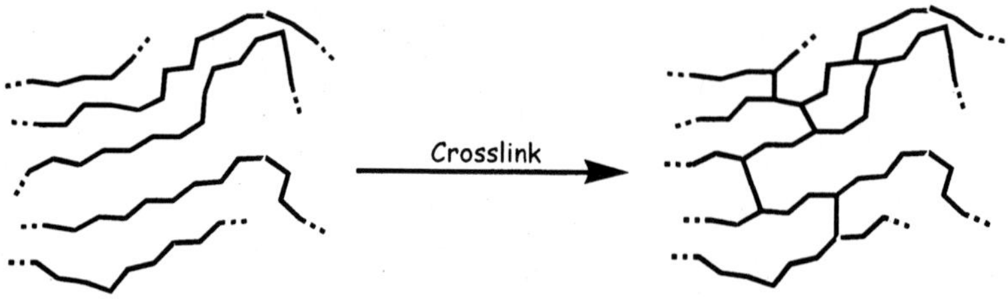

Figure 10. Cross-linking of polymer chains.

Photochromic spiropyrans

In this experiment students create their own indicator of ultraviolet (UV) light exposure, a photochromic spiropyran, and evaluate different sunscreen products to determine their effectiveness in blocking UV light.

What is a photochromic spiropyran? A photochromic substance is one that will change color when irradiated with certain wavelengths of light. A spiropyran is one such molecule (Figure 11). This molecule is colorless until irradiated with UV light. Upon heating, it will turn colorless again. This cycle can be performed over and over. These sorts of molecules are used for many optical applications (*21*). One noteworthy application is in determining how much UV light one is exposed to from sunlight.

Exposure to harmful UV rays exacerbated by ozone depletion causes severe health problems such as skin cancer, cataracts, and impaired immune systems. This radiation can also damage crops and phytoplankton in the ocean. There has recently been a rise in the awareness of the detrimental effects of UV light due to the diminishing ozone layer. This decrease in the protective layer has resulted in large increases in skin cancer among humans (*22*). Short wavelength, high energy (UV-B) rays are more harmful than long wavelength (UV-A) rays, which cause skin to darken. Sunscreens are designed to protect the skin from the dåmaging UV-B rays while allowing UV-A rays to tan the skin. Sunscreens are provided in different SPFs, or skin protection factor ratings; for example, SPF 15 sunscreen provides 15 times longer protection than unprotected skin.

In this experiment students are given a piece of plastic coated with a polymer solution of polymethylmethacrylate (PMMA) and a spiropyran molecule. The PMMA is used as the polymer resin for coating the spiropyran molecule onto the substrate. The spiropyran molecule, which responds to long wavelength UV light, can act as an indicator for UV light exposure. Various sunscreen products with differing SPF levels are coated on the spiropyran indicator sheet, then are exposed to UV light and visualized to determine the effectiveness of the sunscreen. This experiment is an effective way to initiate discussions of global problems such as ozone depletion and to discuss various topics such as color, reversible reactions, and light.

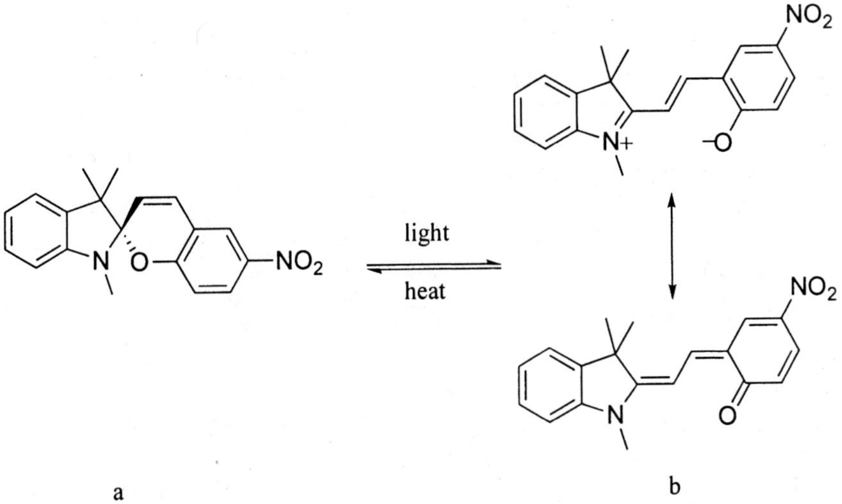

Figure 11. A spiropyran molecule in its colorless (a) and colored (b) form.

K-12 teacher training

Beyond Benign regularly holds training programs for middle school and high school teachers to introduce green chemistry curriculum and teach the teachers how to apply the principles of Green Chemistry to their own classroom procedures. Engaging teachers to help develop the curriculum content and to ensure relevance to a middle or high school classroom enhances the educational expertise of Beyond Benign.

Beyond Benign offers differing levels of K-12 teacher training. A short one-day course involves the introduction of the principles of green chemistry and an introduction to the field. The introduction is followed by hands-on activities such as those described previously. The teachers are left with an understanding of green chemistry and where they can turn for further resources in order to introduce the concepts into a classroom setting. More in-depth training programs are held for middle and high school teacher through two programs: *Solutions in Green Chemistry* (*23*) and *Recipe for Sustainable Science* (*24*). Both training programs are four days long and involve the introduction of the developed curricula, along with exploring methods for "greening" an existing curriculum. Over 400 teachers have been trained through these two programs to date.

Solutions in Green Chemistry is a multi-disciplinary curriculum and teacher-training institute for the high school science classroom. The curriculum was developed by a team of high school teachers and curriculum experts in order to develop a unit that can be introduced into any high school science course. The curriculum is based around a simulation model that engages high school students in the world of science that relates very clearly to their daily lives. The unit encourages invention and inquiry in the study of science and the students learn to solve problems using the new Green Chemistry concepts they have learned. The curriculum is also standards-based to address and demonstrate national standards from multiple science disciplines.

The *Recipe for Sustainable Science* unit introduces Green Chemistry at the middle school level. It is a ten-day interdisciplinary unit containing 46 lesson plans that uses hands-on activities to introduce students to the principles of Green Chemistry. Teachers use a sustainable business model as the mechanism to study Green Science. Students form companies and compete to create a product based upon the company vision they develop. The sustainable pedagogy seeks to explore science from the perspectives of the economy, society and environment in equal measure and to highlight the power and possibilities of scientific discovery to the interdependent nature of these facets of sustainability. Similar to the high school curricula, the *Recipe for Sustainable Science* is also standards-based to introduce material that will fit into required learning for standardized state testing.

The *Solutions in Green Chemistry* and *Recipe for Sustainable Science* units have been developed to reach students of all levels and abilities. The units target those students who have come to think that science is not for them and students to which science has not captured their imaginations. The units are designed to be hands-on and allow students to use their creativity within sound scientific principles. Both units are also available in Spanish for the use in other countries and for the ESL populations in U.S. schools that typically get very little science content in their education. Students will come away from the units with an alternative view of science that relates to their daily lives and teachers will learn how to apply Green Chemistry and sustainability to their own classroom activities.

K-12 curriculum development

Curriculum development at the K-12 level goes hand-and-hand with teacher training. The development of green chemistry curriculum materials must involve collaboration between the green chemistry researchers and the K-12 teacher. Several approaches have been found to be effective in developing curriculum, which can be applied in the K-12 classroom.

Beyond Benign actively engages teachers in the development of K-12 curriculum materials. A team of educators partner with Beyond Benign staff to brainstorm topics and relevant content for a middle school or high school classroom. The team then works together to design content for the appropriate science level. By engaging the stakeholders ahead of time, Beyond Benign can ensure the development of relevant materials.

Another resource, which can aid curriculum materials development, is University students using a service-learning model. Service learning is defined as "a teaching and learning strategy that integrates meaningful community service with instruction and reflection to enrich the learning experience, teach civic responsibility, and strengthen communities." (*25*) Service learning is currently being implemented on College and University campuses across the country.

At the University of Massachusetts Lowell service learning was used to develop curriculum for the 8th grade science classroom. University students were enrolled in a non-science major's science course with a laboratory component. The goal was to develop the curriculum and then obtain feedback from 8th grade science teachers. During the laboratory the University students initially developed 8th grade science curriculum on their own. They were given specific criteria and guidelines for development: 1) the curriculum module must be designed with a green chemistry and sustainability focus, 2) the module must target the state frameworks used to develop standardized exams, 3) the module must have a hands-on laboratory component, and 4) the module must provide ample instruction and explanation to both the student and the teacher. Some of the student modules and their corresponding Massachusetts Framework description are listed in Table I.

The developed curricula has been disseminated to several middle school teachers and feedback will be gathered in order to re-work and reassess the modules to ensure the highest quality content.

Conclusions

We have found green chemistry to be an effective vehicle to foster interest in chemistry and the materials sciences. We see K-12 education and outreach as essential for a sustainable future where green chemistry flourishes and safe, benign products are standard. By forming productive collaborations between researchers and educators at the frontiers of green chemistry, we anticipate an increasing awareness by the general public and K-12 students, of the positive impacts of chemistry on society. Through this new awareness, increased interest in science may then prosper and contribute to the number of students entering the physical sciences. It is, after all, the next generation of students that must answer the much-needed call to make safer materials for a sustainable future.

References

1. *Science and Engineering Degrees: 1966–2004*; NSF 07–307; National Science Foundation, Division of Science Resources Statistics: Arlington, VA, 2006.

Table I. Correlation of Student Modules to Massachusetts Frameworks

Module Title	*MA Framework*	*MA Framework Description*
Energy and Ice Cream	Forms of energy	Differentiate between potential and kinetic energy. Identify situations where kinetic energy is transferred into potential energy and vice versa.
On the "fencity" with density	Properties of Matter	Differentiate between volume and mass. Define density.
How much do you weigh on Saturn?	Properties of Matter	Differentiate between mass and weight recognizing that weight is the amount of gravitational pull on an object.
Chemical Changes vs Physical Changes	Elements, Compounds, and Mixtures	Differentiate between physical changes and chemical changes.
Molecules (That you can eat!)	Elements, Compounds, and Mixtures	Differentiate between an atom (the smallest unit of an element that maintains the characteristics of that element) and a molecule (the smallest unit of a compound that maintains the characteristics of that compound).
What's the deal with mixtures and pure solutions? An experiment with Oobleck	Elements, Compounds and Mixtures	Differentiate between mixtures and pure solutions

2. Heylin, M. Chemistry Grads Post Gains in 2005. *Chem. Eng. News* **2006**, *84* (Jul 24), 43–52.
3. Heylin, M. Chemistry Grads Decline in 2002. *Chem. Eng. News* **2004**, *82* (Mar 29), 48–55.
4. *The Condition of Education 2006*; NCES 2006–071; National Center for Education Statistics: Washington, DC, 2006; Indicator 31.
5. *The Nation's Report Card: Science 2005*; NCES 2006-466; National Center for Educational Statistics: Washington, DC, 2006.
6. Smalley, R. Our Energy Challenge. *Abstr. Pap.—Am. Chem. Soc.* **2003**, *226*, ACS-AICHE.

7. Warner, J. C. Construction of Solar Energy Devices with Natural Dyes. In *Greener Approaches to Undergraduate Chemistry Experiments*; Kirchhoff, M., Ryan, M. A., Eds.; American Chemical Society: Washington, DC, 2002; p 42.
8. Gratzel, M.; O'Regan, B. *Nature* **1991**, *353*, 737–740.
9. Nazeeruddin, M. K.; Kay, A.; Rodicio, I.; Humphry-Baker, R.; Müller, E.; Liska, P.; Vlachopoulos, N.; Grätzel, M. *J. Am. Chem. Soc.* **1993**, *115*, 6382–6390.
10. Nazeeruddin, M. K.; Pechy, P.; Grätzel, M. *Chem. Commun.* **1997**, 1705–1706.
11. Hagfeldt, A.; Grätzel, M. *Acc. Chem. Res.* **2000**, *33*, 269–277.
12. Cannon, A. S.; Guarrera, D. J.; Morelli. A.; Pressler, W.; Warner, J. C. *J. Sol-Gel Sci. Technol.* **2005**, *36*, 157–162.
13. Cannon, A. S.; Warner, J. C.; Saito, K; Trakhtenberg, S.; Whitfield, J. Water Soluble Photocrosslinking Materials in Cosmetics. *Society of Cosmetic Chemists Annual Scientific Seminar Proceedings*, Boston, MA, 2006; 46–47.
14. Lloyd-Kindstrand, L.; Warner, J. C. Thymine-Containing Styrene Polymers as Environmentally Benign Photoresists. *Biopolymers* **2002**, *8*, 165–174.
15. Warner, J. C. Aqueous Photoresists. In *Greener Approaches to Undergraduate Chemistry Experiments*; Kirchhoff, M., Ryan, M. A., Eds.; American Chemical Society: Washington, DC, 2002; p 45.
16. Blackburn, G. M.; Davies, R. J. H. *J. Chem. Soc. C* **1966**, 2239–2244.
17. Lamola, A. A.; Mittal, J. P. *Science* **1966**, *154*, 1560–1561.
18. Farmer, P.; Walker, J. *The Molecular Basis of Cancer*; Wiley Interscience Publishers: New York, 1985.
19. Sancar, A. *Chem. Rev.* **2003**, *103*, 2203–2237.
20. Whitfield, J.; Morelli, A.; Warner, J. C. Enzymatic Reversal of Polymeric Thymine Photocrosslinking with E. coli DNA Photolyase. *J. Macromol. Sci.* **2005**, *A42*, 1541–1546.
21. Tyer, N. W.; Becker, R. S. *J. Am. Chem. Soc.* **1970**, *92*, 1289–1294.
22. U.S. Environmental Protection Agency. Ozone Science: The Facts Behind the Phaseout. http://www.epa.gov/ozone/science/sc_fact.html (accessed Dec 27, 2007).
23. Beyond Benign Foundation. Solutions in Green Chemistry: An Introduction to Green Chemistry in the High School. http://www.beyondbenign.org/outreacheducation/highschool.html (accessed Dec 27, 2008).
24. Beyond Benign Foundation. Recipe for Sustainable Science: An Introduction to Green Chemistry in the Middle School. http://www.beyondbenign.org/outreacheducation/middleschool.html (accessed Dec 27, 2008).
25. National Service-Learning Clearinghouse. http://www.servicelearning.org (accessed Dec 27, 2008).

Chapter 13

Green Chemistry Education: Toward a Greener Day

Mary M. Kirchhoff

Education Division, American Chemical Society, 1155 16th Street, N.W., Washington, DC 20036

Green chemistry began infiltrating the curriculum in the late 1990s, when several enterprising faculty members started introducing greener laboratory experiments, stand-alone courses, and green chemistry modules into their teaching. Widespread coverage of this important topic, however, has been slow to catch on. This is not unusual in education, as curriculum reform is frequently evolutionary rather than revolutionary, and new concepts may take a generation to become embedded within the curriculum. In light of the increasing pressures placed on the planet by humanity, the need to develop a cadre of chemistry professionals who are dedicated to developing and implementing green chemistry practices is more important than ever before. This chapter highlights some of the challenges, opportunities, and strategies for the future in green chemistry education.

Although the concept of green chemistry was introduced in the early 1990s, this approach to the practice of chemistry is still largely absent from the curriculum. The Brundtland Commission defined sustainability in 1987, yet this concept is rarely presented to chemists and chemical engineers during their educational preparation. Faculty members who wish to present chemistry in a global environmental context are hampered by an overcrowded curriculum, few resources that can easily be integrated into existing courses, narrowly focused research problems, perceived lack of rigor, and general inertia. Despite the absence of green chemistry, green engineering, and sustainability from most curricula, the importance of these concepts in the training of scientists and engineers is being echoed by a number of groups:

- "Greater understanding of the societal implications of their work by scientists and engineers will enhance our stewardship of this planet." – *Beyond the Molecular Frontier* (*1*)
- "Chemistry education can only reflect the current practice of chemistry if it also includes the relevance of the discipline through engagement with broader society and the promotion of high ethical standards and environmental performance." – *Exploring the Molecular Vision* (*2*)
- ACS supports "the development of chemistry courses that present the broad scope of modern chemistry, including environmental protection and green chemistry." – *Statement on Science Education Policy* (*3*)

A 2005 report by the National Research Council, *Sustainability in the Chemical Industry* (*4*), highlighted eight grand challenges for the chemical industry in the next 100 years, including the improvement of sustainability education. Internationally, the United Nations has designated 2005–2014 as the Decade of Sustainable Education. This designation offers a unique opportunity to highlight within the curriculum the link between green chemistry and sustainability. A number of organizations in the United States are actively promoting sustainability education, particularly at the collegiate level (*5–8*).

Challenges

Introducing new topics into an already overcrowded chemistry curriculum is not an easy task. Educators may hesitate to introduce green chemistry into their teaching and research because their own background has not adequately prepared them to do so. Identifying and integrating supplementary materials into an existing course can be time consuming, and colleagues may question the value of including green chemistry in the majors' curriculum.

A significant obstacle to the integration of green chemistry and sustainability into the curriculum is the lack of curricular materials that substantively address these topics, though a number of resources have been developed in recent years. For example, several green chemistry texts (*9–11*) are available for use in stand-alone courses. Laboratory manuals (*12, 13*), primarily focused on organic chemistry, enable faculty members to introduce single experiments or an entire curriculum. The *Journal of Chemical Education* has published a number of green chemistry experiments (*14, 15*) as well as other green chemistry-focused papers. Curricular materials are also being disseminated through the University of Oregon's Greener Education Materials (GEMs) online database (*16*). This online resource allows educators to access, contribute, and review green chemistry curricular materials.

Six universities in Germany collaborated to develop NOP (Nachhaltigkeit Organische Chemie Praktikum) Online, an organic chemistry lab course focused on sustainability (*17*). This collection features 75 organic chemistry experiments, a handful of which have been fully evaluated for their environmental impacts.

The American Chemical Society (ACS) offers a number of resources that supplement the curriculum. ACS collaborated with the Royal Society of Chemistry and the Gesellschaft Deutscher Chemiker to produce *Introduction to Green Chemistry* (*18*), a collection of six units that present key concepts in green chemistry. High-school students are the target audience for this resource, whose units incorporate hands-on activities such as a Vitamin C clock reaction. The ACS lab manual *Greener Approaches to Undergraduate Chemistry Experiments* (*13*) is available in both Spanish and Chinese.

Trying to cover too much material in the curriculum is an ongoing problem in terms of presenting new concepts, whether they are green chemistry, nanotechnology, or materials science. The key is replacement, offering a green chemistry example that illustrates a basic concept or reaction type. For example, the aldol reaction is routinely taught in organic chemistry. Providing an example of a solventless aldol reaction (*12*) is a seamless way to introduce students to green chemistry concepts in the context of a classic organic reaction.

The question of rigor is easily addressed by providing real-world examples of green chemistry. Many of the reactions presented to students in the classroom were developed using "brute-force" chemistry—high temperatures and pressures and harsh chemicals. Green chemistry is more subtle; reaction conditions must be carefully chosen to avoid toxic reagents, work at ambient conditions, and minimize the production of waste. Merck received a 2005 Presidential Green Chemistry Challenge Award for its redesigned synthesis of aprepitant, the active ingredient in Emend® (*19*), a drug that is used to treat nausea and vomiting in patients undergoing chemotherapy. The new synthetic pathway dramatically decreases the use of raw materials, water, and energy; eliminates 340,000 L of waste per 1000 kg of product; and nearly doubles the yield when compared with the first-generation synthesis.

Because inertia is a powerful force, a compelling case for green chemistry must be made, and the role of green chemistry in achieving sustainability makes a very strong case. As developing countries become more developed, the global standard of living will rise, leading to increased demand for more consumer goods. Manufacturing these products in a manner that depletes natural resources and generates pollution is not sustainable. The chemical industry is responding by hiring sustainability directors to help guide the environmental stewardship of individual companies. This presents the chemical education community with the opportunity to develop a new generation of chemists who can provide leadership in this area.

Opportunities

Too often, chemistry is taught devoid of context. Chemistry majors can solve complex problems and plan elaborate syntheses, but they don't know why such problems are relevant or to what end such syntheses may be used. Green chemistry provides that context, teaching students that careful design of products and processes can provide environmental, economic, and social benefits—the triple bottom line.

There are a number of options for faculty members to employ in bringing green chemistry into the curriculum. They can integrate green chemistry concepts into the classroom and laboratory, develop stand-alone courses, emphasize green chemistry in their research, or point students in the direction of extracurricular activities such as conferences and workshops.

A long-standing extracurricular activity is the ACS Student Affiliates program, and the American Chemical Society recognizes chapters that participate in green chemistry activities throughout the academic year. Students have hosted speakers, created websites, visited local schools, and organized on-campus poster sessions to raise awareness about green chemistry. This recognition program encourages Student Affiliates chapters to deepen their understanding of green chemistry beyond the classroom.

Green chemistry does not have to be complicated, but it does need to be intentional. Faculty members should look for opportunities to introduce green chemistry concepts and examples into the classroom and the laboratory. A simple example is atom economy, which is a natural fit when discussing the basic types of organic reactions. Addition and rearrangement reactions inherently have a higher atom economy than elimination and substitution reactions. The same concept of atom economy can be introduced into general chemistry in discussions of mass balance.

Laboratory courses offer students the opportunity to experience hands-on green chemistry, and modifications to existing labs can be very effective in communicating green chemistry concepts. In general chemistry, for example, measuring colligative properties is a common laboratory experiment. Aromatic compounds are typically used to measure colligative properties in the teaching laboratory. A variation on this experiment uses fatty acids (*20*), thereby demonstrating the use of safer substances and renewable feedstocks and the conversion of the nontoxic waste stream into a useful product (soap).

The ACS Committee on Professional Training (CPT) has just released its new guidelines for undergraduate chemistry programs (*21*). The new guidelines will afford chemistry departments increased flexibility in offering specific tracks that build on the strengths of the department and individual faculty members. A department could, for example, offer a track in green chemistry, which would provide undergraduates with more in-depth coverage.

There is a potential drawback in offering green chemistry degrees at the undergraduate and graduate level, however. Chemists in other tracks may view environmental protection as the purview of the "green chemists" and fail to realize that these concerns apply to their own research as well. Green chemistry is an approach that is applicable to all areas of chemistry, and it is important that *all* chemists be introduced to these concepts, not just the green chemists.

Strategies for the Future

A major hurdle to introducing green chemistry into the curriculum arises from mainstream chemistry textbooks, few of which contain information on green chemistry. Several market-leading texts (*22–24*) mention green chemistry, but coverage tends to be minimal, with information relegated to sidebars or end-of-chapter boxes. Two ACS texts, *Chemistry in the Community* (*25*) and *Chemistry in Context* (*26*), introduce students to green chemistry concepts. These texts are designed for high school students and undergraduates who are not chemistry majors, respectively, and present chemistry on a need-to-know basis. A more comprehensive approach to green chemistry and sustainability is planned for the seventh edition of *Chemistry in Context.*

Weaving green chemistry through the main body of textbooks at all levels, including general and organic chemistry, is essential for widespread adoption of green chemistry. A review of leading college textbooks revealed that approximately 25 percent make some reference to green chemistry (*27*). Unless green chemistry is more substantively integrated into texts, it is likely to literally remain in the margins of chemistry education.

Green chemistry is often regarded as the province of organic chemistry, even though it is an approach that is applicable to all areas of chemistry. A good collection of teaching resources for organic chemistry exists, but materials for other subdisciplines are in short supply. Practitioners across all areas of chemistry must come to recognize that green chemistry is not an area of specialization, but a strategy that benefits the entire discipline. Accordingly, educators from other subdisciplines who already understand the general applicability of green concepts across all branches of chemistry must be encouraged to develop, use, and disseminate green education materials that relate specifically to their own fields of expertise.

Educators need to become both more familiar with and more comfortable teaching green chemistry concepts, and professional development is an effective means to do so. For example, the University of Oregon's annual summer workshop (*28*) has trained numerous faculty in greener organic chemistry methodologies; as a result, more campuses are adopting greener laboratory experiments.

Both the European Union and the United States host annual summer schools to introduce students to green chemistry concepts and applications. The Venice Summer School will celebrate its 10th anniversary in 2008. This program engages students across Europe in discussions on a variety of topics relevant to green chemistry, such as catalysis, ionic liquids, and alternative reagents. ACS' Summer School on Green Chemistry brings together graduate and postdoctoral students from across the Americas to learn the basics of green chemistry and engineering. Both of these programs provide in-depth exposure to green chemistry, and similar initiatives are needed across the globe to ensure that the rising stars of chemistry and engineering are adequately prepared to institute sustainable practices throughout their careers.

Conclusion

A group of students who participated in ACS' 2005 Summer School on Green Chemistry published an article about their experience in the *Journal of Chemical Education* (*29*). In this article, they referred to the incomplete education they had received because they were not trained in green chemistry:

> "A common belief among the GCSS (Green Chemistry Summer School) participants was that our education would have been significantly enhanced with the incorporation of green chemistry, beginning at the elementary level and continuing throughout graduate course work...Future chemists and chemical engineers must be equipped with the tools necessary to support and promote global sustainability."

These graduate and postdoctoral students recognized that numerous benefits can be realized by introducing green chemistry concepts into the curriculum. Environmentally friendly technologies support student interest in environmental issues, and students recognize that chemistry and environmental protection are compatible. On a practical level, implementing green chemistry in the laboratory can reduce waste and improve safety, yielding both human health and economic benefits. Achieving a sustainable planet will require breakthroughs in green chemistry, all of which begin with advances in green chemistry education.

References

1. *Beyond the Molecular Frontier;* The National Academies Press: Washington, DC, 2003.

2. Society Committee on Education. *Exploring the Molecular Vision,* conference report; June 27–29, 2003, http://portal.acs.org/portal/fileFetch/C/CTP_005564/pdf/CTP_005564.pdf (accessed Sep 30, 2008).
3. American Chemical Society Statement on Science Education Policy, http://portal.acs.org/portal/fileFetch/C/WPCP_007622/pdf/WPCP_007622.pdf (accessed Sep 30, 2008).
4. *Sustainability in the Chemical Industry;* The National Academies Press: Washington, DC, 2006.
5. U.S. Partnership for Education for Sustainable Development, http://www.uspartnership.org (accessed Sep 30, 2008).
6. Higher Education Associations Sustainability Consortium, http://www.heasc.net (accessed Sep 30, 2008).
7. Association for the Advancement of Sustainability in Higher Education, http://www.aashe.org (accessed Sep 30, 2008).
8. Association of University Leaders for a Sustainable Future, http://www.ulsf.org (accessed Sep 30, 2008).
9. Anastas, P. T.; Warner, J. C. *Green Chemistry: Theory and Practice;* Oxford University Press: Oxford, U.K., 1998.
10. Lancaster, M. *Green Chemistry: An Introductory Text;* Royal Society of Chemistry: Cambridge, U.K., 2003.
11. Matlack, A. *Introduction to Green Chemistry;* S. Marcel Dekker: New York, 2001.
12. Doxsee, K. M.; Hutchison, J. E. *Green Organic Chemistry: Strategies, Tools, and Laboratory Experiments;* Thomson Learning: Mason, OH, 2002.
13. *Greener Approaches to Undergraduate Chemistry Experiments;* Kirchhoff, M.; Ryan, M. A., Eds.; American Chemical Society: Washington, DC, 2002.
14. Dintzner, M. R.; Wucka, P. R.; Lyons, T. W. *J. Chem. Educ.* **2006**, *83*, 270-272.
15. Bennett, G.D. *J. Chem. Educ.* **2005**, *82*, 1380-1381.
16. The Greener Education Materials (GEMs) for Chemists Database. http://greenchem.uoregon.edu/gems.html (accessed Sep 30, 2008).
17. NOP Online, http://www.oc-praktikum.de/?lang=en (accessed Sep 30, 2008).
18. Ryan, M. A.; Tinnesand, M. *Introduction to Green Chemistry;* American Chemical Society: Washington, DC, 2002.
19. *The Presidential Green Chemistry Challenge Awards Program: Summary of 2005 Award Entries and Recipients*; U.S. Environmental Protection Agency: Washington, DC; p 6.
20. McCarthy, S. M.; Gordon-Wylie, S. *J. Chem. Educ.* **2005**, *82*, 116-119.
21. *Undergraduate Professional Education in Chemistry: ACS Guidelines and Evaluation Procedures for Bachelor's Degree Programs.* American Chemical Society Committee on Professional Training. *http://portal.acs.org/portal/fileFetch/C/WPCP_008491/pdf/WPCP_008491.pdf* (accessed Sep 30, 2008).

22. McMurry, J. *Organic Chemistry: A Biological Approach*; Thomson Brooks Cole: Belmont, CA, 2006.
23. Brown, T. E.; LeMay, H. E. Bursten, B. E. *Chemistry: The Central Science;* Prentice Hall: Upper Saddle River, NJ, 2006.
24. Solomons, T. W. G.; Fryhle, C. B. *Organic Chemistry*, 8th ed.; John Wiley & Sons, Inc: Hoboken, NJ, 2003.
25. American Chemical Society, *Chemistry in the Community;* W.H. Freeman: New York, 2006.
26. American Chemical Society, *Chemistry in Context;* McGraw-Hill: New York, 2007.
27. Personal communication with Michael Cann, University of Scranton, 2007.
28. Green Chemistry in Education Workshop. http://greenchem.uoregon.edu/Pages/WorkshopApplicationForm.php (accessed Sep 30, 2008).
29. Braun, B.; Charney, R.; Clarens, A.; Farrugia, J.; Kitchens, C.; Lisowski, C.; Naistat, D.; O'Neil, A. *J. Chem. Educ.* **2006**, *83*, 1126-1129.

Indexes

Author Index

Subject Index

D

F

G

H

I

J

K

L

M

N

O

P

Q

R

S

T

U